AF389448

Contemporary Natural Philosophy and Philosophies—Part 2

Contemporary Natural Philosophy and Philosophies—Part 2

Editors

Marcin J. Schroeder
Gordana Dodig-Crnkovic

MDPI • Basel • Beijing • Wuhan • Barcelona • Belgrade • Manchester • Tokyo • Cluj • Tianjin

Editors
Marcin J. Schroeder
Tohoku University
Japan

Gordana Dodig-Crnkovic
Chalmers University of Technology
Sweden
Mälardalen University
Sweden

Editorial Office
MDPI
St. Alban-Anlage 66
4052 Basel, Switzerland

This is a reprint of articles from the Special Issue published online in the open access journal *Philosophies* (ISSN 2409-9287) (available at: https://www.mdpi.com/journal/philosophies/special_issues/Philosophy_and_Philosophies2).

For citation purposes, cite each article independently as indicated on the article page online and as indicated below:

LastName, A.A.; LastName, B.B.; LastName, C.C. Article Title. *Journal Name* **Year**, *Article Number*, Page Range.

ISBN 978-3-03943-535-7 (Hbk)
ISBN 978-3-03943-536-4 (PDF)

Cover image courtesy of Jonathan McCabe.

Contents

About the Editors . . . . **vii**

Marcin J. Schroeder and Gordana Dodig-Crnkovic
Contemporary Natural Philosophy and Philosophies—Part 2
Reprinted from: *Philosophies* 2020, 5, 22, doi:10.3390/philosophies5030022 1

Richard de Rozario
Matching a Trope Ontology to the Basic Formal Ontology
Reprinted from: *Philosophies* 2019, 4, 40, doi:10.3390/philosophies4030040 7

Ronald B. Brown
Breakthrough Knowledge Synthesis in the Age of Google
Reprinted from: *Philosophies* 2020, 5, 4, doi:10.3390/philosophies5010004 19

Andreas Stephens and Cathrine V. Felix
A Cognitive Perspective on Knowledge How: Why Intellectualism Is
Neuro-Psychologically Implausible
Reprinted from: *Philosophies* 2020, 5, 21, doi:10.3390/philosophies5030021 33

Cathrine V. Felix and Andreas Stephens
A Naturalistic Perspective on Knowledge How: Grasping Truths in a Practical Way
Reprinted from: *Philosophies* 2020, 5, 5, doi:10.3390/philosophies5010005 47

Johannes Schmidl
De Libero Arbitrio—A Thought-Experiment about the Freedom of Human Will
Reprinted from: *Philosophies* 2020, 5, 3, doi:10.3390/philosophies5010003 59

Cristian S. Calude and Karl Svozil
Spurious, Emergent Laws in Number Worlds
Reprinted from: *Philosophies* 2019, 4, 17, doi:10.3390/philosophies4020017 65

Roman Krzanowski
What Is Physical Information?
Reprinted from: *Philosophies* 2020, 5, 10, doi:10.3390/philosophies5020010 79

Gordana Dodig-Crnkovic
Natural Morphological Computation as Foundation of Learning to Learn in Humans, Other
Living Organisms, and Intelligent Machines
Reprinted from: *Philosophies* 2020, 5, 17, doi:10.3390/philosophies5030017 97

Joseph E. Brenner and Abir U. Igamberdiev
Philosophy in Reality: Scientific Discovery and Logical Recovery
Reprinted from: *Philosophies* 2019, 4, 22, doi:10.3390/philosophies4020022 113

Marcin J. Schroeder
Contemporary Natural Philosophy and Contemporary *Idola Mentis*
Reprinted from: *Philosophies* 2020, 5, 19, doi:10.3390/philosophies5030019 159

About the Editors

Marcin J. Schroeder is a Specially Appointed Professor at the Global Education Center of the Institute for Excellence in Higher Education at Tohoku University, Sendai, Japan, and Professor Emeritus at Akita International University, Akita, Japan. He holds a Ph.D. degree in Mathematics from Southern Illinois University at Carbondale, USA, and an M.Sc. degree in Theoretical Physics from the University of Wroclaw, Poland. He is the founding Editor-in-Chief of the journal Philosophies and President of the International Society for the Study of Information (IS4SI). In his more than forty years of work in several universities of Europe, America, and Asia, he has taught courses in mathematics, logic, statistics, physics, and information science. His research in several disciplines outside of mathematics has adapted to his multiple roles in the academic community to include topics such as development and implementation of university curricula, intercultural communication, and international education, but the primary theme of his more recent publications is the theoretical, mathematical study of information and symmetry and the related philosophical reflection on these concepts.

Gordana Dodig-Crnkovic is Professor in Computer Science at Mälardalen University and Professor of Interaction Design at Chalmers University of Technology in Sweden. She holds Ph.D. degrees in Theoretical Physics from the University of Zagreb and in Computer Science from Mälardalen University. Her research is in morphological computation and the study of information, as well as information ethics and ethics of emergent technologies. She has a long record of teaching courses in theory of science and research methodology, transdisciplinary research methods, computational thinking, formal languages, professional ethics, and research ethics, among others. She published the books *Investigations into Information Semantics and Ethics of Computing* (2006), *Information and Computation Nets* (2009), six edited volumes (2007, 2011, 2013, 2017, 2019, and 2020) with Springer and World Scientific, and one edited book with MDPI. She is past President of the Society for the Study of Information, member of the editorial board of the World Scientific Series in Information Studies and Springer SAPERE series, and a member of the editorial board of six journals.

 *philosophies*

Editorial

Contemporary Natural Philosophy and Philosophies—Part 2

Marcin J. Schroeder [1,*] and **Gordana Dodig-Crnkovic** [2,3]

[1] Global Learning Center, Tohoku University, Sendai 980-8576, Japan
[2] Department of Computer Science and Engineering, Chalmers University of Technology, 412 96 Gothenburg, Sweden; gordana.dodig-crnkovic@chalmers.se
[3] School of Innovation, Design and Engineering, Mälardalen University, 722 20 Västerås, Sweden
* Correspondence: mjs@gl.aiu.ac.jp

Received: 10 September 2020; Accepted: 10 September 2020; Published: 14 September 2020

Abstract: This is a short presentation by the Guest Editors of the series of Special Issues of the journal *Philosophies* under the common title "Contemporary Natural Philosophy and Philosophies" in which we present Part 2. The series will continue, and the call for contributions to the next Special Issue will appear shortly.

Keywords: natural philosophy; philosophy of nature; naturalism; unity of knowledge

1. Introduction

The present Special Issue is Part 2 of the project of a series under the common title *Contemporary Natural Philosophy and Philosophies*, intended as a venue for publishing the results of research and philosophical reflection seeking a unified vision of reality threatened by the centrifugal forces of growing specialization. In the Introduction to Part 1, the project was presented within the mission of the journal *Philosophies*: "From the Philosophies program [1], one of the main aims of the journal is to help establish a new unity in diversity in human knowledge, which would include both 'Wissen' (i.e., 'Wissenschaft') and 'scīre' (i.e., 'science'). 'Wissenschaft' (the pursuit of knowledge, learning, and scholarship) is a broader concept of knowledge than 'science', as it involves all kinds of knowledge, including philosophy, and not exclusively knowledge in the form of directly testable explanations and predictions. The broader notion of scholarship incorporates an understanding and articulation of the role of the learner and the process of the growth of knowledge and its development, rather than only the final product and its verification and validation. In other words, it is a form of knowledge that is inclusive of both short-term and long-term perspectives; it is local and global, critical and hypothetical (speculative), breaking new ground. This new synthesis or rather re-integration of knowledge is expected to resonate with basic human value systems, including cultural values" [2].

We would also like to give place in this modern Natural Philosophy to the human in the natural world, both as an active subject and as an integral part of nature, for whom the world comes as an interface [Ref. Otto Rössler, Endophysics: The World as an Interface]. Separation between human and nature, thought and feeling, rational and intuitive, knowledge how and knowledge that embodiment and abstraction, physical and mathematical relations to the world have led to the compartmentalization of human reality in various non-communicating domains. All should have place in this new synthetic network of knowledges under the umbrella of Contemporary Natural Science, in which there is place for the whole human and natural world in coexistence and co-creation.

Part 1 brought an excellent collection of 23 articles (listed in Appendix A) addressing the ideas of the revitalization, revival or recreation of Natural Philosophy, naturalization of some domains of inquiry, or presenting examples of research carried out in the spirit of Natural Philosophy published in book format [3]. Part 2 was announced simply as a continuation of this effort, and we are happy and proud to present the next collection of 10 contributions.

2. Contributions to Part 2

The contributions offered in Part 2 are steps toward a better understanding of reality and of the way to carry out the inquiry so that we have a broader perspective without losing the unified vision or striving even for achieving a higher level of unification. Even though the papers constituting Part 2 were written independently, upon close inspection, we can identify a thread which binds them into a discourse on the relation between knowing (episteme) and existence (ontos) sometimes as if one paper was intended as a response to another.

The two entirely independent papers, *Matching a Trope Ontology to the Basic Formal Ontology* by Richard de Rozario and *Breakthrough Knowledge Synthesis in the Age of Google* by Ronald B. Brown address the central issue of Contemporary Natural Philosophy of the unification of knowledge in the practical context of the present information technology [4,5].

The first of the papers *Matching a Trope Ontology to the Basic Formal Ontology* by Richard de Rozario is focused on applied ontology which, as the author writes "is as much philosophy as engineering". The main difference between the applied and original form of ontology in the theoretical aspect of the former is that it frequently restricts its attention to more specific conceptual frameworks of particular knowledge disciplines. It is justified by its engineering function to provide methods of combining large databases and developing software working with such combination. From this point of view, applied ontology gives practical tools for the implementation of the goals of Contemporary Natural Philosophy and can also serve as its quasi-empirical field.

The subject of this paper is the relation between Basic Formal Ontology (BFO) and trope ontology. It provides a logical matching, identifies key ontological issues that arise, and concludes with general observations about the matching, such as that matching of universals is generally straightforward, but not the matching between relations. The issues addressed in the paper seem analogical to those which appear in the ontological discussions of the forms of realism, in particular, in the context of structural realism.

The second paper, *Breakthrough Knowledge Synthesis in the Age of Google* by Ronald B. Brown could be, by analogy to the first, understood as applied epistemology studying a web-based knowledge synthesis method relevant in today's information technology environment with its easy access to online interactive tools and an expansive selection of digitized peer-reviewed literature. The paper offers an innovative method of synthesis based on a grounded theory methodology to organize, analyze, and combine concepts from an intermixed selection of quantitative and qualitative research, inferring an emerging theory or thesis of new knowledge. We can find, in the conclusion of the paper, the statement "Breakthrough knowledge has been shown to occur most often when prior knowledge is mixed with current knowledge." The two papers have basically similar goals, but they differ in the focus on ontological, respectively epistemological issues.

There are two other papers in the collection of Part 2, this time by the same authors but with possibly different roles in research or writing indicated by the reversed order of names which present studies of knowledge from the perspective of neuro-psychology. While the papers reported before had, as their objectives, fostering the integration of human knowledge by the use of information technology, the two papers reported now are more oriented towards the understanding of knowledge as purely human capacity.

The first of the papers *A Cognitive Perspective on Knowledge How: Why Intellectualism Is Neuro-Psychologically Implausible* by Andreas Stephens and Cathrine V. Felix defends the thesis of the fundamental distinction between "knowledge how" and "knowledge that" and provides empirically backed refutation of the claims (for instance of Stanley-style intellectualism) that knowledge how can be reduced to knowledge that [6]. Moreover, the authors demonstrate that the distinction leads to neuro-psychologically plausible understanding of knowledge.

The second paper *A Naturalistic Perspective on Knowledge How: Grasping Truths in a Practical Way* by Cathrine V. Felix and Andreas Stephens is on a similar subject of the distinction between knowing how and knowing that, this time in the elaborated perspective of the practical consequences of knowledge [7]. The authors argue that a plausible interpretation of cognitive-science input concerning knowledge—even if one accepts that *knowledge how* is partly propositional—must involve an element of knowing how to act correctly upon the proposition, and this element of knowing how to act correctly cannot, itself, be propositional.

The paper *De Libero Arbitrio—A Thought-Experiment about the Freedom of Human Will* by Johannes Schmidl, although at first sight is on a very different subject of free will, has an indirect affinity to the previously reported paper in its focus on action [8]. The former paper relates knowing how to human action which, as the authors demonstrate, makes it distinct from knowing that. The paper reported now is about the issue of the distinction between human action understood as an expression of the free will and behavior determined in biological or physiological terms which of course reduces the role of knowledge how to biological process which can be described in propositional form, i.e., knowing that. Schmidl defends irreducibility using a thought-experiment which demonstrates that a subjective consciousness can break any forecast about its physical state independently of the method of its detection, which refutes the claims about its purely deterministic role. The thought-experiment picks up on an idea of the philosopher Alvin I. Goldman.

The paper *Natural Morphological Computation as Foundation of Learning to Learn in Humans, Other Living Organisms, and Intelligent Machines* by Gordana Dodig-Crnkovic adds yet another dimension and a wider context to the study of knowledge not restricted anymore to its form of exclusively human capacity [9]. Dodig-Crnkovic considers the integrative view of the natural, the artificial, and the human-social knowledge and practices. She gives the learning process a central role for acquiring, maintaining, and managing knowledge, both theoretical and practical. The paper explores the relationships between the present advances in understanding of learning in the sciences of the artificial (deep learning, robotics), natural sciences (neuroscience, cognitive science, biology), and philosophy (philosophy of computing, philosophy of mind, natural philosophy).

Dodig-Crnkovic explores the question about the inspiration from nature, specifically its computational models such as info-computation through morphological computing for the development of machine learning and artificial intelligence, and the question about how much, on the other hand, models and experiments in machine learning and robotics can motivate, justify, and inform research in computational cognitive science, neurosciences, and computing nature.

The central idea of the potential contribution to the design for a system to reach human-level intelligence can be understanding of the mechanisms of 'learning to learn' as a step towards deep learning with symbolic layer of computation/information processing in a framework linking connectionism with symbolism.

The topic of the computational aspects of knowledge in natural, human, and artificial systems can be associated with another paper *What Is Physical Information?* by Roman Krzanowski [10]. However, this author's approach gives priority to the physicalist view of information expressed in his attempt to separate the two forms of information qualified as abstract and physical with a focus on the latter, which he describes as information which has an objective existence, a lack of meaning, and which can be quantified. These three features he identifies as characteristics of physical phenomena. In Krzanowski's view, physical information can be expressed as an organization of natural or artificial entities which corresponds to syntactic information, with no function of representing the world (carrying meaning). He also argues that concepts of (abstract) information that are associated with meaning also depend (to a substantial degree) on physical information, in the same way as semantic information in computing is built upon a given syntax.

The traditional approach making the sharp distinction between the physical (expressed as formal) and abstract (expressed as representational) aspects of reality presented in Krzanowski's paper is in strong contrast to the following three papers. The strongest contrast is with the position presented in the paper *Spurious, Emergent Laws in Number Worlds* by Cristian S. Calude and Karl Svozil, which is an excellent example of the innovative methodology fitting the needs of Contemporary Natural Philosophy [11]. Calude and Svozil refer to Heidegger's Fundamental Question of Metaphysics "Why is there anything at all, rather than nothing?" as their point of departure, but their work goes not in the direction of ontology but, rather, epistemology. The paper has been mostly concerned with the formal consequences of existence under the least amount of extra assumptions. Heidegger's existent (physis, nature, or world) is, here, a World Number—a real number presented in the binary form. Calude and Svozil consider philosophical precedents of this view (in the past considered as a plurality, i.e., world built of numbers, not a number), but they are of secondary importance in this work.

Although this might not have been their intention, their view can be associated with Kripke semantics and the concept of possible worlds. Each possible world consists of all mutually, logically consistent true propositions about the world. Then, the actual world is one of the possible worlds. All propositions about the world bound into a conjunction into one proposition which can be numerically encoded into an infinite binary string (for instance, using Gödel's numbers) and we arrive at Calude and Svozil's Number World without sacrificing much of the traditional philosophical convictions. It seems that the authors' way of thinking is guided more by algorithmic considerations than logical, but in either interpretation, the Number World is convincing. However, the next step is more difficult. They consider the patterns within the Number World as natural laws. However, these laws (patterns) are not intrinsic. The analysis carried out by the authors is from the outside, God-like position which comes with the combinatorial methods (e.g., Ramsey theory).

Whether this perspective is acceptable or not for the reader, epistemological consequences it brings are impressive: "As it turns out, existence implies that an intrinsic and sophisticated mixture of meaningful and (spurious) patterns—possibly interpreted as 'laws'—can arise from xáos. The emergent 'laws' abound, they can be found almost everywhere. The axioms in mathematics find their correspondents in the 'laws' of physics as a sort of 'lógos' upon which the respective mathematical universe is 'created by the formal system'. By analogy, our own universe might be, possibly deceptively and hallucinatory, be perceived as based upon such sorts of 'laws' of physics".

The paper *Philosophy in Reality: Scientific Discovery and Logical Recovery* by Joseph E. Brenner and Abir U. Igamberdiev gives, in some sense, a competing view [12]. While the Number World of Calude and Svozil is governed by classical logic of language (or, if someone prefers, logic of Turing-type computing), Brenner and Igamberdiev make a clear distinction between epistemological logic of language and Logic in Reality. They propose a sublation of linguistic logics of objects and static forms by a dynamic logic of real physical–mental processes designated as the Logic in Reality (LIR). In their generalized logical theory, dialectics (logical reasoning) and semiotics are recovered from reductionist interpretations and reunited in a new synthetic paradigm centered on meaning and its communication. Their theory constitutes a meta-thesis composed of elements from science, logic, and philosophy.

The last paper *Contemporary Natural Philosophy and Contemporary Idola Mentis* by Marcin J. Schroeder presents his rather idiosyncratic but minimally restrictive view of Contemporary Natural Philosophy to set the stage for a critical review of habits of thought, which can be detected in the present attempts to achieve goals of Contemporary Natural Philosophy within existing scientific and philosophical methodology [13]. Following Baconian tradition, the habits are grouped in the three (non-exclusive and non-exhaustive) categories of the Idols of the Number, the Idols of the Common Sense, and the Idols of the Elephant. Once again, the examples of the idols in the paper can be understood as unintentional critical responses to some papers in Part 2. For instance, the paper provides arguments against the overestimated distinction between the quantitative and qualitative methods of science (as one of the Idols of the Number). One of the Idols of Common Sense is the common misconception regarding the role of definitions obscuring the process of comparison of competing theories based on different conceptual frameworks. Finally, one of the Idols of the Elephant is the tendency to "flatten" the vision of reality.

Without any intention on the side of editors or contributors the collection can be viewed as a dialog between several distinct, complementary, but also cooperating views on reality and on the ways of its inquiry and the role of humans in this context.

Funding: The editorial work on this SI received no external funding.

Acknowledgments: The Guest Editors would like to express their gratitude to the authors who contributed to this Special Issue and to numerous anonymous peer reviewers whose work helped in improving the quality of published contributions.

Conflicts of Interest: The authors declare no conflict of interest.

Appendix A

The List of Contributions to Volume 1 of Contemporary Natural Philosophy and Philosophies available at https://www.mdpi.com/books/pdfview/book/1331:

Gordana Dodig-Crnkovic and Marcin J. Schroeder, *Contemporary Natural Philosophy and Philosophies.*

Bruce J. MacLennan, *Philosophia Naturalis Rediviva: Natural Philosophy for the Twenty-First Century.*

Nicholas Maxwell, *We Need to Recreate Natural Philosophy.*

Stanley N. Salthe, *Perspectives on Natural Philosophy.*

Joseph E. Brenner, *The Naturalization of Natural Philosophy.*

Andrée Ehresmann and Jean-Paul Vanbremeersch, *MES: A Mathematical Model for the Revival of Natural Philosophy.*

Arran Gare, *Natural Philosophy and the Sciences: Challenging Science's Tunnel Vision.*

Chris Fields, *Sciences of Observation.*

Abir U. Igamberdiev, *Time and Life in the Relational Universe: Prolegomena to an Integral Paradigm of Natural Philosophy.*

Lars-Göran Johansson, Induction and Epistemological Naturalism.

Klaus Mainzer, *The Digital and the Real Universe. Foundations of Natural Philosophy and Computational Physics.*

Gregor Schiemann, *The Coming Emptiness: On the Meaning of the Emptiness of the Universe in Natural Philosophy.*

Koichiro Matsuno, *Temporality Naturalized.*

Robert E. Ulanowicz, *Dimensions Missing from Ecology.*

Matt Visser, *The Utterly Prosaic Connection between Physics and Mathematics.*

Kun Wu and Zhensong Wang, *Natural Philosophy and Natural Logic.*

Lorenzo Magnani, *The Urgent Need of a Naturalized Logic.*

Roberta Lanfredini, *Categories and Dispositions. A New Look at the Distinction between Primary and Secondary Properties.*

Rafal Maciag, *Discursive Space and Its Consequences for Understanding Knowledge and Information.*

Harald Atmanspacher and Wolfgang Fach, *Exceptional Experiences of Stable and Unstable Mental States, Understood from a Dual-Aspect Point of View.*

Włodzisław Duch, *Hylomorphism Extended: Dynamical Forms and Minds.*

Robert Prentner, *The Natural Philosophy of Experiencing.*

Robert K. Logan, *In Praise of and a Critique of Nicholas Maxwell's In Praise of Natural Philosophy: A Revolution for Thought and Life.*

References

1. Schroeder, M.J. The Philosophy of Philosophies: Synthesis through Diversity. *Philosophies* **2016**, *1*, 68. [CrossRef]

2. Dodig-Crnkovic, G.; Schroeder, M.J. Contemporary Natural Philosophy and Philosophies. *Philosophies* **2018**, *3*, 42. [CrossRef]

3. Dodig-Crnkovic, G.; Schroeder, M.J. (Eds.) *Contemporary Natural Philosophy and Philosophies—Part 1*; MDPI: Basel, Switzerland, 2019. Available online: https://www.mdpi.com/books/pdfview/book/1331 (accessed on 10 September 2020).
4. De Rozario, R. Matching a Trope Ontology to the Basic Formal Ontology. *Philosophies* **2019**, *4*, 40. [CrossRef]
5. Brown, R.B. Breakthrough Knowledge Synthesis in the Age of Google. *Philosophies* **2020**, *5*, 4. [CrossRef]
6. Stephens, A.; Felix, C.V. A Cognitive Perspective on Knowledge How: Why Intellectualism Is Neuro-Psychologically Implausible. *Philosophies* **2020**, *5*, 21. [CrossRef]
7. V. Felix, C.; Stephens, A. A Naturalistic Perspective on Knowledge How: Grasping Truths in a Practical Way. *Philosophies* **2020**, *5*, 5. [CrossRef]
8. Schmidl, J. De Libero Arbitrio—A Thought-Experiment about the Freedom of Human Will. *Philosophies* **2020**, *5*, 3. [CrossRef]
9. Dodig-Crnkovic, G. Natural Morphological Computation as Foundation of Learning to Learn in Humans, Other Living Organisms, and Intelligent Machines. *Philosophies* **2020**, *5*, 17. [CrossRef]
10. Krzanowski, R. What Is Physical Information? *Philosophies* **2020**, *5*, 10. [CrossRef]
11. Calude, C.S.; Svozil, K. Spurious, Emergent Laws in Number Worlds. *Philosophies* **2019**, *4*, 17. [CrossRef]
12. Brenner, J.E.; Igamberdiev, A.U. Philosophy in Reality: Scientific Discovery and Logical Recovery. *Philosophies* **2019**, *4*, 22. [CrossRef]
13. Schroeder, M.J. Contemporary Natural Philosophy and Contemporary *Idola Mentis*. *Philosophies* **2020**, *5*, 19. [CrossRef]

 *philosophies*

Article

Matching a Trope Ontology to the Basic Formal Ontology

Richard de Rozario

The Hunt Lab for Intelligence Research, The University of Melbourne, Parkville, VIC 3010, Australia;
richard.derozario@unimelb.edu.au

Received: 19 May 2019; Accepted: 15 July 2019; Published: 18 July 2019

Abstract: Applied ontology, at the foundational level, is as much philosophy as engineering and as such provides a different aspect of contemporary natural philosophy. A prominent foundational ontology in this field is the Basic Formal Ontology (BFO). It is important for lesser known ontologies, like the trope ontology of interest here, to match to BFO because BFO acts like the glue between many disparate ontologies. Moreover, such matchings provide philosophical insight into ontologies. As such, the core research question here is how we can match a trope ontology to BFO (which is based on universals) and what insights such a matching provides for foundational ontology. This article provides a logical matching, starting with BFO's top entities (continuants and occurrences) and identifies key ontological issues that arise, such as whether universals and mereological sums are equivalent. This article concludes with general observations about the matching, including that matching to universals is generally straightforward, but not so much the matching between relations. In particular, the treatment of occurrences as causal chains is different in the trope ontology, compared to BFO's use of time arguments.

Keywords: ontology; BFO; tropes; applied philosophy

---

1. Introduction

The field of applied ontology came to prominence in the 1990s [1,2], driven by knowledge engineering issues. In order to achieve coherent and shareable knowledge bases, engineers and scientists sought answers to questions in essence like "what is the meaning of a physical quantity?" [3] or "what exactly constitutes a gene?" [4]. The similarity to questions raised in philosophical ontology was readily apparent [5], opening up the possibility of joint engineering and philosophical ontology research.

Arguably, applied ontology is not merely the application of philosophy *to* other disciplines like engineering. Rather, the discipline also informs philosophy itself—at the very least by raising new questions about the existence of things, but hopefully also by novel approaches and answers to those questions. As such, applied ontology is one pathway in the journey of reconstituting a natural philosophy, in the sense of Dodig-Crnkovic and Schroeder's connected conceptual engineering [6].

The interaction with philosophical ontology is especially visible at the foundational ontology level. Foundational ontologies[1] deal with fundamental aspects of our world, such as "material objects", "events", "being part of" and so forth. They underpin ontologies of particular domains such as biomedicine or information systems [2,7] (pp. 115–139).

Currently, one of the most prominent foundational ontologies is the Basic Formal Ontology (BFO) [8]. BFO is a realist ontology, based on a relatively small set of universals and relations[2]. That is, BFO asserts that the (real) universe can be carved up into universals such as *objects*, *processes*,

---

[1]   Also called "upper ontologies".
[2]   BFO includes relations (as per the specification), but these are as yet not included in the bfo.owl file. The discussion here refers to the specification.

boundaries and *qualities*. By "universals", BFO means " … what all members of a natural class or natural kind such as a *cell*, or *organism*, or *lipid*, or *heart* have in common … not only in the realm of natural objects such as enzymes and chromosomes, but also in the realm of material artifacts such as flasks and syringes, and also in the realm of information artifacts such as currency notes and scientific publications" [9] (p. 13). BFO has been successful in supporting engineering and scientific research in areas such as genetics, information systems and defence [10].

In previous work on the ontology of competitive intelligence [11], I developed a core ontology based on particularized relations, or *tropes*, which are fundamentally different from the universals that BFO is based on (discussed in detail below). It is beyond the scope of this article to review the arguments for and against tropes (of which there are many flavors), suffice it to say that tropes are prominent in ontological theory that addresses the nature and quantity of properties (like being "a chair" or "being red") [12]. At the applied level, tropes as particularized relations provide a way to connect causes to the structural relations that define entities, providing a seamless foundation for both entities and causality [11]. From either perspective, universals vs. tropes is a core ontological commitment.

The previous work compared the trope ontology to several other ontologies, but not BFO. Given the prominence of BFO, it is important for lesser known ontologies such as the trope ontology to provide at least some comparison, but preferably a "matching" to BFO. This is important because foundational ontologies often act as the "glue" to connect disparate domain ontologies and, if a matching is possible, it would link the trope ontology to the wider world of ontologies based on BFO. Moreover, given the subject matter of foundational ontologies, discussion of such matchings informs philosophical ontology—albeit with a constructive, rather than critical emphasis.

Matching is commonly defined as "the process of finding relationships or correspondences between entities of different ontologies" [13] (p. 39). I include in this looking for points of similarity, where terms in one ontology can be coherently defined in terms of another. As such, "matching" reflects a perspectivist stance towards ontology. That is, it recognizes that different ontologies may reflect different aspects or viewpoints of reality (including perhaps some level of denial of that reality).

In summary, this article examines how the trope ontology was matched to BFO and what ontological insights we might obtain from such a matching. The next section will briefly outline the trope ontology and the remaining sections will match core terms of the trope ontology to BFO. In the conclusion, I will provide some general observations about the results of the matching and the relationship between the two ontologies.

2. The Trope Ontology

Given the relative obscurity of the trope ontology, I will provide a brief overview here. For more detail, see my previous research [11].

The trope ontology is grounded in a view that relations between entities are real and exist primarily as individual relations, rather than universal relations. For example, when John stands next to Jane, it is not just John and Jane who exist, but also the relation of John's standing next to Jane. If John is also standing next to Jake, then that "standing next to" has a distinct existence from the "standing next to" of John and Jane. Moreover, these individual "standing next to" relations cannot exist without the entities that they bind. This strong dependence of relations on their relata supports calling such entity/relation constructions "tropes", although it represents only one view of tropes [12]. Philosophically, I have defended tropes by following Armstrong's reasoning [14] towards particularized universals, but rejecting Armstrong's argument that "states of affairs" are needed to bind relations to objects [11] (pp. 29–30). In this way, particularized relations (tropes) become an alternative theory that underpins relational realism. At the applied level, this results in individually identifiable relations, which provide a very convenient way to implement change and causal relations (discussed below).

Methodologically, I try to adhere to parsimony and minimize the different kinds of relations that are admitted into the ontology, if for no other reason than the effort required to carefully examine each relation for coherence in the ontology. On the other hand, one typically tries to choose relations that are as expressive as possible—i.e., relations with which one can say or represent as much as possible. So, the key principle of parsimonious expressiveness (or "say the most with the least") guides much of the work in the trope ontology. As such, the trope ontology is based on two main kinds of relations: mereological parthood and a primitive causality relation that ranges over the parthood relations. The semantics and formal properties of these relations are as follows.

The two primitive relations are represented with a simple predicate schema:

$$p(N, part, X, Y), \tag{1}$$

$$p(M, cause, A, B), \tag{2}$$

Here, N and M represent unique identifiers for each relation, typically constructed as a list of numerals[3]. The second argument in the predicate is the kind of relation (e.g., part or cause) and the remaining arguments represent the entities that the relation binds.

A small digression is needed on the meaning of "kind of relation" in the context of trope theory. At first glance, it seems that "kind" introduces universals again. For instance, each individual parthood relation seems to instantiate a universal parthood. Might this not undermine the supposed "fundamental difference" (asserted in the introduction) between the trope ontology and a universal-based ontology like BFO? The short answer is "no", for the following reasons. Firstly, these are not the universals you are looking for. BFO distinguishes between universals (i.e., what members of certain classes of entities have in common) and relations such as "instantiates" or "has participant" [8] (p. 7). There is a sense in which universals in BFO are more complex entities (e.g., *cells, flasks* or *currency notes*) than the comparatively bare relations. Secondly, even if we admitted that relations have characteristics like universals, it does not oblige a trope ontologist to commit to relations *as universals*. There are two common realist options other than universals [12]. One is to posit resemblance as primitive, so that tropes are *resembling tropes* without having to distinguish "resemblance" from a relation like parthood. An alternative option is to posit a higher order resemblance relation, which avoids vicious regress by supervening on "lower level" particulars. The trope ontology is based on the former (i.e., *resembling tropes* as primitives) as it is the simpler of the two. In case "simplicity" seems like an inadequate justification, it should be noted that a choice of "primitives" needs to be accepted, at some level, in any ontology. Moreover, primitives are preferred where they reduce the number of ancillary ontological commitments needed (i.e., commitments that are needed only to maintain coherence, rather than to do the main "definitional" ontology work). Arguably, this is based on no more than a philosophical stance away from an "overpopulated universe [that] is in many ways unlovely" [15], but I shall nevertheless let it rest on that.

The specific kinds of relations and their arguments can be expanded for particular domain ontologies, but the core of the trope ontology has at least the parthood and causal relations. The parthood relation is formally defined through the axioms of General Extensional Mereology [16] (pp. 31–37). Note that I interpret general sums here as any collection of individuals under some meaningful relation, and not arbitrary collections [11] (p. 73), leaving the discussion of "meaningful" to individual cases.[4] Furthermore, it is important to understand parthood in a general sense in the trope ontology. For example, John's hand is part of him and therefore part of John's family, even though the hand is arguably part of John in a different way than John's being part of his family. However, the meaning of parthood is understood in the context of the relata of *hand* and *family*. This approach follows Eschenbach and Heydrich [17] by combining a general mereology with restricted domains.

---

[3]   The use of lists of numerals as identifiers helps with the inference of transitive relations. For example, if we have p([1], part, a, b) and p([2], part, b, c), then we can infer p([1,2], part, a, c), where [1,2] represents the unique identifier for the a-c parthood relation.

[4]   This stipulation supports a matching to the universals, discussed later.

The causal relation is transitive, irreflexive, anti-symmetric and ranges over parthood relations (and possibly other relations, if such relations exist within a specific domain ontology that is based on the trope ontology) [11] (pp. 86–96). The justification of causal primitives follows the reasoning of causal singularists like Richard Taylor [18], who argue that causality like "the fire started because of the lightning" has a stronger connection than mere constant conjunction (like "the fire started and there was lightning"). Moreover, singularists would also argue that the laws or rules that are often posited as underpinning causal relations merely describe the very causal relations that exist in the world [11] (pp. 86–90). In the trope ontology, we end up with a primitive causal relation that ranges over other relations—in particular over the individual parthood relations. The causal relation in the trope ontology should be understood as a "simple" relation that merely represents a linkage between an antecedent situation and its consequential situation in terms of the primitive relations (e.g., parthood) that the causal relation ranges over. This means that, in contemporary language, relations that are not ordinarily thought of as "causes" may indeed be causes in the ontology. For example, if John intentionally moves from Sydney to Melbourne, then both his intention and his prior being in Sydney are causal antecedents to his being in Melbourne. The way that John's being in Sydney is a causal antecedent to his being in Melbourne is similar to Aristotle's material cause, whereas John's intention is more aligned with a final cause [19] (94b.3).

The causal relations are *replacement* relations, in that the consequence replaces the antecedent. For example, if "John's being part of Sydney" causes "John's being part of Melbourne" then "John's being part of Sydney" no longer exists[5] at the end of that causal chain. This also applies in cases where arguably the antecedent could persist. For example, if "John's desire to be in Melbourne" causes him to be in Melbourne, it might be that his desire persists even when he is in Melbourne. Insofar as the causal relation reflects a change between situations, we must decide that either relations persist from the antecedent by default, or that they perish by default. The physical world indicates that antecedents are replaced by their consequent, therefore relations perish by default. For example, the relation of "John being in Sydney" perishes in the causal process. However, this implies that persistent antecedents must be explicitly renewed as additional consequences. Thus, if John's desire causes him to be in Melbourne and his desire persists, then, in order to persist, his antecedent desire must also cause (i.e., renew) his further desire. There is a certain level of representational choice in this part of the ontology, because one could assert persistence as the default representation (with explicit assertion of perishing relations). For example, in the case where John's walking and his intention to walk cause him to (keep) walking, one might argue that the physical situation persists by default. However, it seems that linguistically at least, we tend to interpret physical relations as perishing under causation, and this is what the trope ontology orients on.

The trope ontology is further extended with support for multiple worlds, represented with additional structure in the identifiers. For example, p([w:1], part, john, kitchen) might represent the relation that john is in the kitchen in world w (where w stands for some identifier). These multiple worlds form the basis for a modal description of the ontology, where we can talk of "possibility" and "necessity" [11] (pp. 108–114). In this case, I use the causal relation as the *accessibility relation* between worlds [20]. For instance, if John currently is in one room of a house, then it is possible that John can be in another room by asserting that John's being in one room (in the current world) is cause for John's being in another room (in another world). The stipulation of different worlds enables us, in this case, to interpret "cause" modally as "may cause". Note that using the causal relation as the basis of accessibility means that the ontology can only support a modal logic that stops short of *normal modal logic* ("S5" in Kripke semantics [20]). Rather, the modal logic that is supported is S4 plus anti-symmetry.

Worlds also support the representation of informational states of entities. That is, the content of such an informational state—i.e., the *information*—can be represented with the same kinds of predicates

---

[5] "exists" in the sense of John's location in the physical world.

as used for the rest of the ontology, but isolated in their own world. For example, we might represent that John is thinking about being in Melbourne as,

p([0:1],part, thinking(1), 'John').
p([1:10], part, 'John', 'Melbourne').

The representation uses "reserved" functional terms such as thinking (1) to refer to an informational state, the *content* of which is represented by the predicates with the world indicator (1). Predicates with different world indicators are isolated from each other. It is only the informational state (in this case, "thinking") as a whole that exists as part of John, not the relations within the informational state. This isolation is necessary to avoid that whatever John thinks also automatically exists in the world outside his thoughts. So, just because John *thinks* he is in Melbourne, does not mean he *is* in Melbourne—his thinking and the physical universe are two different worlds. Assertion of informational states provides the pathway for defining core elements of the sociotechnical domain ontology that the trope ontology was first concerned with. That is, agentive "intentions" are defined as information states of some entity (i.e., the agent). Those information states (e.g., the state of neurons, or the magnetic state of electronic memory) may cause subsequent situations in the world.

The trope ontology is not only particularist in its view of relations, but also particularist in scope and intent. That is, trope ontologies will typically be limited to particular domains or investigations. The aim is not that such ontologies are complete in their own right but can be used as modules in a network of ontologies. Moreover, there is allowance for enhancements or even corrections of ontologies. As such, ontological inferences on a certain domain ontology that is based on the trope ontology will necessarily be limited to that domain ontology. To put it another way, inferences for a particular ontology do not necessarily hold when that ontology is changed, or when other ontologies are added. However, that limitation also enables us to make a simplifying[6] "closed world" assumption for inferences on the ontology. That is, any inferences are particular to a specific ontology and do not extend beyond unless explicit linking is asserted.

The trope ontology was intended as a foundation for constructive exploration of ontology, where one adds elements to the ontology based on particular cases. That is, in exploring the ontology of a domain, one starts by attempting to describe particular cases or examples of situations or states of affairs in that domain. Such attempts will highlight the kinds of entities and relations that an ontology needs to represent. The core of the trope ontology provides the scaffolding of basic relations that is suggestive of how as yet undefined terms or relations might work. For example, we might begin investigating an ontology of migration by trying to express John's intended move between cities. With parthood, causation and informational states we might attempt a simple version like, "John is part of Melbourne, because John was a part of Sydney; and John intended that John will be a part of Melbourne." So far, the example ontology is severely incomplete. That is, entities like John and the cities are captured as undifferentiated entities in the ontology. Similarly, we are using a generic relation like parthood only to capture the general idea of what is happening in this case, but we would likely need more details to capture the semantics of "existing in a city". However, it is exactly these attempts at definition with a minimal ontology that reveal where more definition is needed, and thus supporting an iterative process of construction, inferential testing, and elaboration of the ontology.

At the practical level, the trope ontology is implemented as a logic program, with utilities to convert to other formats. In particular, programs are available to convert to and from "controlled natural language" statements [21]. Sentences like the one above (John's move) can be entered as plain text and are then converted to collections of predicates that comprise the ontology.

---

6   Simplifying in the sense of making one closed world assumption for the ontology, rather than asserting for example the scope of each class. An overall closed world assumption also enables efficient inference in certain logics, such as the logic programming language Prolog.

3. From Tropes to Universals

In BFO, "an entity is anything that exists. BFO assumes that entities can be divided into instances (your heart, my laptop) and universals or types (*heart, laptop*)" [8] (p. 6). In other words, for BFO, a type like *heart* is as real as the numerous hearts that instantiate it. The trope ontology is also realist in its stance. However, in the trope ontology your heart's relationship to your body has a distinct existence from Jane's heart's relationship with her body. In contrast, BFO views the property of your heart being a *part of* your body (at some time) as the exact same "part of" property as Jane's heart being *part of* her body[7]. Given this difference between the trope ontology and BFO, we now face the question of how to match the different accounts of reality that BFO and the trope ontology present.

There are two problems here: one is how to match tropes to general relations and, secondly, how to match certain particularized tropes to the universals of BFO. Note that we are just aiming for matching one ontology to another and not a justification of one ontology versus the other. However, a matching, if successful, will hopefully underpin a philosophy of reconciliation on these points.

Mereology, the theory of parts and wholes, provides a way to get from particular relations to general relations. That is, we could match general relations, such as *part of* used in BFO, to sums (collections) of all the particular tropes (such as your heart being part of your body and John's heart being part of his body) used in the trope ontology[8]. What matching does in this case is to acknowledge that in BFO there exists a *general* relation, but such a general relation does not exist in the trope ontology—and yet, wherever BFO's *part of* is applied, a *part of trope* can be understood to exist (from a tropist viewpoint). This is not an extensional equivalence, because for every *a being part of b*, BFO counts three things (*a, b* and *part of*)[9], whereas the trope ontology counts only one (*a being part of b*). However, we can match one such trope to a corresponding BFO structure of *a being part of b*, without having to compromise either ontology's philosophical principles (i.e., without having to commit to unacceptable entities in either ontology).

A similar use of mereological sums also provide a partial pathway to universals. The general extensional mereology discussed above allows for the existence of any collection of entities under some relation. For instance, in the trope ontology, objects would exist as the collection of individual objects, on the assumption that there is a particularized relation of " ... is an object" for every individual object. If we assume that tropes like "x is an object" exist, then the matching to BFO's universal *object* can proceed in the same way as the matching of relations like "part of". However, many such "relations" (e.g., BFO universals such as *object, process, quality*) do not exist in our sparse trope ontology. On the other hand, we can easily assert in the trope ontology that "x is part of objects", by using the existing parthood relations. What is missing for matching purposes is the assertion that there is a relation *in virtue of which* x is part of objects. This situation can be solved by simply asserting the existence of such relations. Note that we wo not necessarily add the relations as primitives to our ontology. Rather, for certain sums, we add an additional assertion that there exists a relation under which the individuals become a sum—without necessarily defining that relation in a formal sense. In essence, these are "placemarker" assertions. We'll use the following schema to assert the existence of such covering relations, nominating the functional term "universal" as a reserved term for this purpose:

$$p([n], \text{part}, \text{universal}(X), \text{universal}(X)).$$

For example, in BFO the universal *object* is defined as a "maximal causally unified material entity". In the trope ontology, if my watch is asserted as an object, we would use the following predicates to represent this:

---

[7] BFO version 2 would define such a property more precisely as continuant_part_of, but our point remains the same.

[8] We have to take care not to fall afoul of self-referential regress, but will not address that detail here.

[9] Actually, it is not entirely clear whether BFO counts "part of". The specification says such relations are not first-class citizens (i.e., entities), but does not say what that means in terms of existence.

p([1], part, 'my watch', objects).

p([2], part, universal(object), universal(object)).

The reflexive declaration of the universal *object* simply establishes the existence of the relation under which the individuals (like my watch) are part of the mereological sum of objects. As such, there is an unstated, but implied and axiomatic inference from the universal (noted as a singular term, *object*) to the corresponding mereological sum (noted as a plural term, like *objects*). The rule can be schematically stated as follows.

$$p(\_, part, universal(X'), universal(X')) \longrightarrow p(\_, part, X, \_) \text{ or } p(\_, part, \_, X). \tag{3}$$

where X' is the singular form of X[10].

In other words, the trope ontology keeps track of the extension of the universal through the corresponding mereological sum. Thus, the transitive property of the subtype relation of the universal is inferable through the transitivity of parthood relations. For instance, a further assertion that "An object is a material entity", would add the following relations to the ontology:

p([3], part, objects, material-entities).

p([4], part, universal(material-entity), universal(material-entity)).

It can be readily seen that:

- my watch is part of objects,
- objects are part of material-entities
- and there is an (implicit) rule matching objects and material-entities to their respective universals

Therefore, it also the case that, by transitivity of the subtype relation, my watch belongs to the universal *material-entity*.

These assertions of the existence of special relations enables us to consistently match BFO's universals with mereological sums in the trope ontology. Note that BFO, at least in its OWL 2 implementation, uses *classes* in a similar way to keep track of the transitive properties of universals [22].

The account so far enables us to match (mereological) tropes to corresponding general relations and universals. Next, we need to look at the specific universals that BFO commits to.

4. Accounting for Continuants and Occurrents

BFO carves entities into *continuants*, which retain their identity over time, and *occurrents*, which have temporal parts. For example, John (a particular human) remains himself over time (i.e., John is a continuant), but John's life (an occurrent) has different parts from time to time. That does not mean that John exists all the time. John can *exist at* a particular time, but whenever John exists, then he exists entirely as John. On the other hand, John's life has different parts in each time period: he is born, he eats a meal, he has a birthday party, etc. Each of these experiences is an *occurrent part of* his life. To complete this picture, it may also be the case that John (the continuant entity) has different parts at different times. For example, he may have a beard today, but be clean shaven tomorrow—so, something may be *part of* a continuant *at a time*. Occurrents like processes or events may also have parts, but we do not need to specify "at some time", because occurrents come with time built-in, so to speak. In other words, a process stretches over time and any part of that process occupies a fragment of that time. Lastly, we make the connection between continuants and occurrents by allowing an occurrent to *have a participant*. So, John participates in John's life.

---

[10] Note that the distinction between plural and singular is not formally necessary, but a naming convention maintained for consistency with BFO's naming convention of universals as singular.

By contrast, the trope ontology asserts a primitive causal relation that may exist between entities, or more precisely between relations of entities[11]. So, John's life comprises the sum of causally linked stages of John, where those stages are defined by individual relations. Note that in the trope ontology, a continuant can be the mereological sum of all its parts, without the causal separation of those parts. For example, there may be a relation of "John having a natural right hand" and another of "John having an artificial right hand", with a causal link between them (where one relation replaces the other). In this case, if we *ignore* the causal link, then the mereological sum of John (i.e., the continuant) has both a natural right hand, as well as an artificial right hand—but that seemingly strange sum in the trope ontology only exists when one "sums up" without regard for the causal relations. This overly broad summing up would be tantamount to saying, "what are all the right hands that John has, regardless of time".

The existence of an occurrent without regard of time cannot match to BFO's view of an occurrent existing in its entirety through time. The reason is that parthoods that occur at particular times in BFO are always specified with the temporal designator (e.g., "John's hand is part of John at time t"). As such, one can only get a sum of *John at a certain time*, ensuring the impossibility of a sum of John with *and* without an artificial hand in BFO. Given this disparity in views of what constitutes the sum of John, the matching from the trope ontology to BFO should either exclude parts that are impermanent, or always attach a temporal property to parts.

To illustrate the preceding treatment of events as named causal sequences, consider the following sentences (to avoid noisy punctuation, we are using lowercase names for entities):

john is part of melbourne, because john was part of sydney;

and john intended that john will be part of melbourne.

The process from john was part of sydney upto john is part of melbourne is the move of john.

These sentences translate into predicate form in the ontology text as follows (where terms like "the move of john" are transformed into functions like *move(john)* and we assume the existence of certain universals):

p([5], part, john, sydney).

p([1], part, john, melbourne).

p([30:32], part, john, melbourne).

p([31], part, intending(30), john).

p([33], cause, [5], [1]).

p([34], cause, [31], []).1

p([35], part, 'cause*'[[5]],[[1]]), move(john)).

p([37], part, move(john), moves).

p([39], part, universal(move), universal(move)).

Predicate [35] represents the notion that the causal chain is part of an occurrence called the "move of john" (i.e., *move(john)* in functional form). We use a special predicate called 'cause*' to indicate that the first argument refers to a chain of relations starting with predicate [5] and ending with predicate [1]. Once the chain of events is a "named" entity, that entity can then be treated like any other entity. For example, the last two predicates reflect that the "move of john" is part of the mereological

---

[11] Keeping in mind that tropes are an integral complex of a relation and the entities it relates.

sum of *moves*. In other words, John's move is a *move*, where *move* is a universal that ultimately would be an *occurrent* in BFO's terms.

Note that the trope ontology does not use a time argument in its occurrences. Rather, causal chains are demarcated by the events they span. One can calculate a time metric based on the lengths of chains, but we can speak about occurrences entirely without needing explicit time parameters. This means that matching from the trope ontology causal chains to BFO occurrence membership is straightforward, but not so much for matching to participation or parthood relations that have time parameters. At the minimum, we would need to pick a time reference point, but we might also need an explicit causal relation in BFO (if that could be countenanced). Moreover, causal chains allow different time metrics along different chains (think of the mind experiment of the astronaut on a fast extra-terrestrial trip who experiences a different length of time than people on earth). In other words, while we can match to occurrence universals in a straightforward way, the matching of corresponding relations depends on the specific scope of the ontology.

On the other hand, occurrents as causal chains does provide for a straightforward way to implement process boundaries and regions. That is, the relations that demarcate the causal chain can also be used to demarcate process boundaries. The further matching of regions would mirror the matching of immaterial entities that is discussed in the next section.

5. Other Universals

BFO divides continuants into independent, specifically dependent and generically dependent continuants. Independent continuants are either material or immaterial entities, of which the former include common objects like tables, elephants and the like. The latter divides into *continuant fiat boundaries* and *sites*, which support the ontology of things like holes and geographic boundaries [8] (p. 41). Following the process discussed above, all these can be matched straightforwardly from the trope ontology to their respective universals in BFO. For example,

HMAS-Beagle is part of Sydney-harbour.

The hold of HMAS-Beagle is part of HMAS-Beagle and a hold.

A hold is a continuant-fiat-boundary.

The example would be represented by the following predicates, readily seen to be following the same matching principles for universals described above.

p([1], part, 'HMAS-Beagle', 'Sydney-harbour').

p([3], part, hold('HMAS-Beagle'), 'HMAS-Beagle').

p([4], part, hold('HMAS-Beagle'), holds).

p([2], part, universal(hold), universal(hold)).

p([5], part, holds, 'continuant-fiat-boundaries').

p([7], part, universal('continuant-fiat-boundary'), universal('continuant-fiat-boundary')).

One aspect about such matchings does need clarification. Namely, the trope ontology would need extra predicates to capture the rules that are part of BFO. For example, material entities cannot be part of immaterial entities. Since the trope ontology is primarily expressed as a logic program (i e , Prolog), it is a relatively simple matter to add extra predicates that capture such rules. One approach is to implement the rules as constraints that can be checked for an entity. For example, below is a possible constraint on any entity that if it is part of *immaterials*, then it cannot also be a part of *materials*.

constraint(X) ⟵ part(X, immaterial), not(part(X, material)).

Specifically, dependent continuants are either a *quality*, such as "the mass of this piece of gold" or "John's being the biological son of Jane", or a *realizable entity*, which are either a *role* or a *disposition*. Realizable entities are like qualities but are not always part of the entity that they adhere to. For example, "John's ability to sleep" is a disposition, because John does sleep, but not always (assuming John is an ordinary person). One can see the same with roles, which are dependent continuants that an entity has due to circumstances and which leave the bearing entity physically unchanged if the role is removed [8] (p. 58). For example, John's being CEO would be a role. *Generically dependent continuants* are like their specific cousins but can be copied between bearers. For example, the arrangement of chess pieces on a board can be exactly copied to another chess board.

The implementation of such dependent continuants is structurally simple through the use of functional terms. Previous examples, such as "the hold of HMAS-Beagle" already show the use of functional terms. Dependent continuants such as qualities would use functional terms as well. For example, "the temperature of 37-C is part of john" in predicate form would be p([1], part, temperature('37-C'), john). However, qualities such as these are only part of their bearers in the most general interpretation of parthood. That is, we can only say in the most general sense that John's temperature is also part of the mereological sum of all humans (of which John is part). We will probably want more nuance to reflect the nature of dependency. The trope ontology already has an example of generic dependence in the case of informational states. There, we use reserved terms to indicate the special nature of the entity. That in turn enables particular treatment (i.e., logical rules) of informational states. On that foundation of informational states, the trope ontology has previously defined roles as entities in some description (i.e., information) that classify entities in the world [11] (p. 182). However, such an approach is more nominal than in BFO, because it relies on the entities with the information for the role to exist. Nonetheless, the same approach of special terms to represent qualities can be used. For example, we might preface qualities with "quality of", as in "the quality of the temperature of 37 °C". This would enable us to recognize qualities for what they are and assure correct inferential rules. This approach follows Eschenbach and Heydrich's use of restricted domains, mentioned earlier.

6. Conclusions

The trope ontology is oriented towards building up an ontology from individual cases, rather than focusing on universals. Universals, in the sense of entities that form collections under some covering relation, emerge in the trope ontology as entities that need to be talked about as collections and need to be distinguished for some reason from other entities. This reflects more of a "bottom up" approach than BFO. However, it also comprises an approach of developing ontologies of bounded scope. Limiting the ontological scope is something that the trope ontology has in common with BFO.

Matching universals in BFO to corresponding entities in the trope ontology is simple, as long as we formally recognize the existence of universals in the first place (i.e., as covering relations of mereological sums). From that point onwards, being a universal in the trope ontology is essentially a property that certain mereological sums have. If a universal can be wholly defined in terms of parthood (or possibly with additional relations that also exist in a trope ontology for a specific domain), then the universal can be defined in the trope ontology directly using the corresponding covering relation of a mereological sum. Alternatively, the existence of the universal can be asserted without formal definition of its trope relation, perhaps with annotation of the informal definition, as is the case with various universals in BFO.

The matching between the primitive causal relations in the trope ontology and *occurrences* in BFO is more complicated. What is available in the trope ontology is the ability to declare causal chains as entities, where those entities match BFO's occurrences. However, temporal measures like "time" are secondary inferences in the trope ontology. For example, a point in time would be derived from the length of a causal chain. Therefore, while causal chains can be classed as occurrences, some relations like "being part of X at time t" require specific assumptions about the measurement of time. Moreover, in the trope ontology a continuant is the mereological sum of all its parts, regardless of causal relations.

This is not the case in BFO, because some mereological relations are only specified with explicit time arguments. As such, the matching of continuants that fall under causal relations will need to exclude causally dependent parts.

In general, the trope ontology can relatively simply match to universals in the Basic Formal Ontology, because universals have an extension that can be represented by corresponding individuals in the trope ontology, and because the trope ontology allows for mereological sums under some covering relation, where that covering relation matches the universal. However, since the trope ontology has causal relations that are not part of BFO, the matching of occurrence properties and relations is not as straightforward. With regard to occurrences, the trope ontology can be seen as an ontology of different granularity that can be matched with BFO, but where the matching requires assumptions about the nature of time.

Funding: This research received no external funding.

Acknowledgments: My thanks to Tim van Gelder for helpful review comments. My thanks also to the anonymous reviewers for the many insightful comments.

Conflicts of Interest: The authors declare no conflict of interest.

References

1. Smith, B. Applied Ontology. A New Discipline is born. *Philos. Today* **1998**, *29*, 5–6.
2. Chandrasekaran, B.; Josephson, J.; Benjamins, V. What are ontologies, and why do we need them? *IEEE Intell. Syst.* **1999**, *14*, 20–26. [CrossRef]
3. Gruber, T.R. Toward principles for the design of ontologies used for knowledge sharing? *Int. J. Hum. Comput. Stud.* **1995**, *43*, 907–928. [CrossRef]
4. Ashburner, M.; Ball, C.A.; Blake, J.A.; Botstein, D.; Butler, H.; Cherry, J.M.; Davis, A.P.; Dolinski, K.; Dwight, S.S.; Eppig, J.T.; et al. Gene Ontology: Tool for the unification of biology. *Nat. Genet.* **2000**, *25*, 25–29. [CrossRef] [PubMed]
5. Smith, B.; Welty, C. Ontology: Towards a new synthesis. In Proceedings of the Formal Ontology in Information Systems (FOIS'01), Ogunquit, ME, USA, 17–19 October 2001; Volume 10, pp. 3–9.
6. Dodig-Crnkovic, G.; Schroeder, M. Contemporary Natural Philosophy and Philosophies. *Philosophies* **2018**, *3*, 42. [CrossRef]
7. Keet, C.M. *An Introduction to Ontology Engineering*; University of Cape Town: Cape Town, South Africa, 2018.
8. Smith, B.; Almeida, M.; Bona, J.; Brochhausen, M.; Ceusters, W.; Courtot, M.; Dipert, R.; Goldfain, A.; Grenon, P.; Hastings, J. *Basic Formal Ontology 2.0: Specification and User's Guide*; National Center for Ontological Research: Buffalo, NY, USA, 2015.
9. Arp, R.; Smith, B.; Spear, A.D. *Building Ontologies with Basic Formal Ontology*; MIT Press: Cambridge, MA, USA, 2015.
10. Smith, B. Basic Formal Ontology Users. Available online: https://basic-formal-ontology.org/users.html (accessed on 18 June 2019).
11. De Rozario, R. Competitive Intelligence: An Ontological Approach. Ph.D. Thesis, University of Melbourne, Melbourne, Australia, 2009.
12. Maurin, A.-S. Tropes. In *The Stanford Encyclopedia of Philosophy*; Zalta, E.N., Ed.; Metaphysics Research Lab, Stanford University: Stanford, CA, USA, 2018.
13. Euzenat, J.; Shvaiko, P. *Others Ontology Matching*; Springer: Berlin/Heidelberg, Germany, 2007; Volume 18.
14. Armstrong, D.M. *Universals: An Opinionated Introduction*; Westview Press: Boulder, CO, USA, 1989.
15. Quine, W.V. On what there is. *Rev. Metaphys.* **1948**, *2*, 21–38.
16. Simons, P. *Parts: A Study in Ontology*; Clarendon Press: Oxford, UK, 1987.
17. Eschenbach, C.; Heydrich, W. Classical mereology and restricted domains. *Int. J. Hum. Comput. Stud.* **1995**, *43*, 723–740. [CrossRef]
18. Taylor, R. Causal Power and Human Agency. In *Introduction to Metaphysics: The Fundamental Questions*; Schoedinger, A.B., Ed.; Prometheus Books: Amherst, NY, USA, 1973; pp. 326–340.
19. Aristotle. Aristotle Physics. In *The Basic Works of Aristotle*; Random House: New York, NY, USA, 2001.

20. Garson, J. Modal Logic. In *The Stanford Encyclopedia of Philosophy*; Zalta, E.N., Ed.; Metaphysics Research Lab, Stanford University: Stanford, CA, USA, 2018.
21. De Rozario, R. The Trope Ontology. Available online: https://github.com/RdR1024/tropeontology (accessed on 5 December 2019).
22. Overton, J. *BFO Repository Including Source Code and Latest Documents: BFO-Ontology/BFO*; Basic Formal Ontology (BFO): New York, NY, USA, 2019.

 *philosophies*

Perspective

Breakthrough Knowledge Synthesis in the Age of Google

Ronald B. Brown

School of Public Health and Health Systems, University of Waterloo, Waterloo, ON N2L 3G1, Canada;
r26brown@uwaterloo.ca

Received: 25 January 2020; Accepted: 28 February 2020; Published: 4 March 2020

Abstract: Epistemology is the main branch of philosophy that studies the nature of knowledge, but how is new knowledge created? In this perspective article, I introduce a novel method of knowledge discovery that synthesizes online findings from current and prior research. This web-based knowledge synthesis method is especially relevant in today's information technology environment, where the research community has easy access to online interactive tools and an expansive selection of digitized peer-reviewed literature. Based on a grounded theory methodology, the innovative synthesis method presented here can be used to organize, analyze and combine concepts from an intermixed selection of quantitative and qualitative research, inferring an emerging theory or thesis of new knowledge. Novel relationships are formed when synthesizing causal theories—accordingly, this article reviews basic logical principles of associative relationships, mediators and causal pathways inferred in knowledge synthesis. I also provide specific examples from my own knowledge syntheses in the field of epidemiology. The application of this web-based knowledge synthesis method, and its unique potential to discover breakthrough knowledge, will be of interest to researchers in other areas, such as education, health, humanities, and the science, technology, engineering, and mathematics (STEM) fields.

Keywords: knowledge synthesis; epistemology; breakthrough knowledge; domain-specific knowledge; web-based search; grounded theory; Bradford Hill criteria; association; causation; mediation

---

1. Introduction

Great minds throughout human history have endeavored to understand the nature of knowledge, which forms the subject of the branch of philosophy known as epistemology, a word that translates to understanding [1]. In his dialogue Theaetetus, written around 369 BC, the philosopher Plato originated a definition of knowledge as justified true belief [2]. Despite criticism by epistemologists and other philosophers, such as Gettier in 1963 [3], Plato's contribution to the definition of knowledge has endured to this day. But how is new knowledge created, especially in other branches and areas of philosophy, such as logic, education, and science? Sir Isaac Newton wrote in 1678, "If I have seen further it is by standing on ye shoulders of giants" [4]. Newton's statement implies that he discovered new knowledge by building upon prior knowledge discovered by others. More recently, researchers examining more than 28 million studies and over 5 million patents discovered that breakthroughs in almost all fields of knowledge are more likely to occur as large amounts of prior knowledge are mixed with current, extant knowledge, confirming Newton's observation [5].

The definition of breakthrough means to overcome a barrier, and breakthrough knowledge implies overcoming barriers to advance knowledge. One barrier to knowledge advancement throughout history has been a powerful status quo which resists novel ideas. For example, while conducting his scientific research at the University of Padua, Italy, Galileo Galilei complained in a letter to scientist Johannes Kepler in 1610 that "these philosophers shut their eyes to the light of truth" [6].

The philosopher Thomas Kuhn [7] later described how most advances in scientific knowledge involve incremental changes within the conventional paradigm. Kuhn further noted how scientific revolutions occur periodically as new knowledge breaks through conventional barriers and causes a disruptive shift in the reigning paradigm. Another historical barrier to the advancement of knowledge occurred throughout Europe in the Middle Ages, when recorded information was owned exclusively by elite sectors of society, usually the clergy and members of academia. As knowledge spread with the advent of the printing press, the power of the Catholic Church was reformed, and the printing press had an influential effect on the Renaissance and the Scientific Revolution [8]. Since then, modern society has witnessed a relentless movement toward the democratization of the public's access to knowledge, especially in the age of digital technology.

Using today's web-based interactive tools such as Google's ubiquitous search engine and online databases, students, educators, practitioners, research scientists and inventors have an unprecedented opportunity to discover breakthrough knowledge by synthesizing current and prior knowledge available online. As academic libraries have digitized much of their content, no longer must students, practitioners, and researchers descend into the dark and dusty basements of institutional buildings seeking microfiches of archived literature. And yet, despite advances in accessing information online, a coming revolution in breakthrough knowledge appears to lie beyond the horizon. Students seeking new knowledge may feel hopelessly overwhelmed as they are bombarded with an overload of redundant online information [9], much of it of questionable veracity. In a quest to discover breakthroughs, research scientists may lack the advantage of leveraging online information search tools [5], inhibiting their capacity to step outside their disciplines and generate innovative, novel theories with the potential to produce a revolutionary paradigm shift in scientific concepts and practices [7]. The aim of this perspective article is to introduce concepts of researching and writing a web-based knowledge synthesis, which is intended to empower researchers across diverse disciplines with the skills to discover breakthrough knowledge in the age of Google.

2. Synthesis

The traditional use of the word art means skill [10]. The art or skill of knowledge synthesis relies on the ability to retrieve information from peer-reviewed studies and form an academic literature review, which provides more than an annotated bibliography of summaries [11]. A synthesis, the assembly of parts into a new whole, organizes and interprets the concepts, connections, controversies and constraints of a body of literature, filling in gaps and generating new insights, perspectives, directions and novel explanations about the research topic. Table 1 lists over two-dozen types of knowledge synthesis methods [12].

Table 1. Knowledge synthesis methods.

Knowledge Synthesis Method	Author
Bayesian meta-analysis	Sutton [13]
Content analysis	Stemler [14]
Critical interpretive synthesis	Dixon-Woods, [15]
Cross-design synthesis	Droitcour [16]
Ecological triangulation	Banning [17]
Framework synthesis	Pope [18]
Grounded theory	Strauss [19]
Interpretive synthesis/Integrative synthesis	Noblit, [20]
Meta-ethnography	Noblit [20]
Meta-interpretation	Weed [21]
Meta-narrative	Greenhalgh [22]
Meta-study	Paterson [23]
Meta-summary	Sandelowski [24]

Table 1. *Cont.*

Knowledge Synthesis Method	Author
Meta-synthesis	Sandelowski [25]
Mixed studies review	Pluye [26]
Narrative review/summary	Dixon-Woods [27]
Narrative synthesis	Popay [28]
Qualitative cross-case analysis	Yin [29]
Qualitative meta-synthesis	Jensen [30]
Qualitative systematic review/evidence synthesis	Grant [31]
Quantitative case survey	Yin [32]
Realist review/synthesis	Pawson [33]
Textual Narrative synthesis	Lucas [34]
Thematic analysis	Mays [35]
Thematic synthesis	Thomas [36]

Note: Table based on Kastner et al. [12].

Other terms for knowledge synthesis include research synthesis, evidence synthesis, and scientific synthesis. Syntheses can integrate current knowledge to inform policy and best practices, as in systematic reviews and meta-analyses, or syntheses can create new knowledge by combining evidence from a wide variety of sources [37]. The latter type of synthesis for new knowledge generation at the source of the flow of scientific information, from theory to practice, is the topic of this perspective article. The term emerging synthesis has been used to describe the ongoing evolution of newer types of synthesis method in which a wide diversity of quantitative and qualitative findings, data, and research designs are combined together to contribute new knowledge and theory to a research area [38].

Explanatory theories that explain causes and effects are qualitatively different from descriptive theories that categorize, organize, and describe phenomena [39]. Hjørland's domain-analysis is an example of a descriptive theory used in library science to organize knowledge according to the specific contents of information within a knowledge domain [40]. I propose that differences between explanatory and descriptive theories are similar to differences between explanatory and descriptive knowledge syntheses. For example, in addition to its use in systematic reviews and meta-analyses, descriptive knowledge syntheses categorize and organize information in taxonomies, ontologies, encyclopedias, databases, library systems, and complex networks. Furthermore, data mining methods have been used to predict new information in complex networks, such as the prediction of paired protein interactions in a protein database, and the prediction of interactions within social networks [41]. However, these methods have limitations. Interactive predictions in social networks, based on observed structural patterns, were shown to have a low rate of correctness [42]. In addition, the method for predicting protein interactions, based on classification of interaction types, does not explain how predicted interactions function, which must be determined with follow-up studies [43], On the other hand, explanatory knowledge syntheses logically combine concepts to infer new explanatory theories and hypotheses which may lead more directly to new explanatory knowledge.

3. Synthesis in Education

Knowledge synthesis is not only an important theoretical concept in the philosophy of knowledge; it plays a role in the philosophy of science, where it also serves as a practical research method, and it is a valued concept as well in the philosophy of education. Teaching the skills necessary for researching and writing knowledge syntheses is an important educational objective. The ability to synthesize material begins as an academic skill developed in secondary school and post-secondary school [44]. Bloom's original 1956 taxonomy of educational objectives included synthesis, the ability to assemble parts into a new whole, along with other educational objectives, such as knowledge, comprehension, application, analysis, and evaluation [45]. A revised version of Bloom's taxonomy, Figure 1, advanced synthesis to the highest educational level as part of Creating [46].

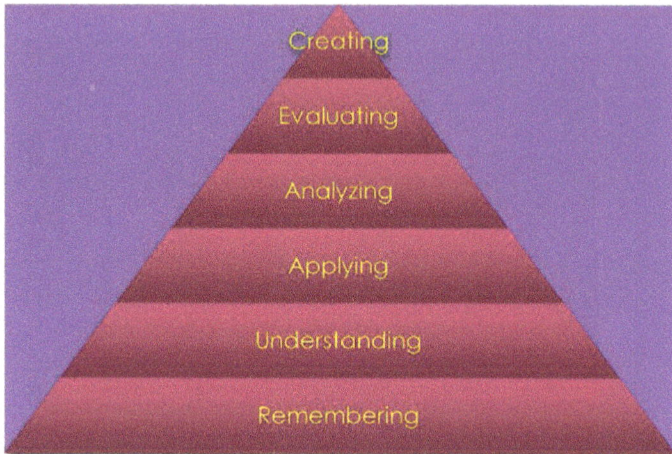

Figure 1. Bloom's taxonomy. Based on Forehand [46].

Training in synthesis skills should be acquired early in a researcher's career [47]. Today's evidence-based medicine is increasing the need for biomedical students to acquire research skills in information retrieval, critical judgment, statistical analysis, and writing [48]. Writing with rigorous analysis and logic is a critical professional skill required by most businesses, industry [49], and government agencies [50]. The National Commission on Writing in America's Schools and Colleges recommends that teachers encourage students to view writing as an enjoyable method for learning and discovery [51]—synthesizing new knowledge is a writing skill that can fulfill that recommendation by stimulating students with the excitement of exploration and discovery, leading to potential breakthrough knowledge. However, methods of synthesis writing must be developed to help students acquire proficiency in the skills of selecting, organizing, and associating information.

Recently, a method improved synthesis writing skills in students by employing note taking for information selection, and providing students with a graphic matrix organizer that presented information side-by-side to more easily draw associations between texts [52]. This method of teaching synthesis writing could be combined with use of web-based interactive tools in the classroom, which have been shown to enhance student engagement and improve learning experiences [53]. For example, web search engines could be used to teach students how to search, select, and synthesize online text sources in subject areas of interest to them. In addition to advancing keyboard, language and writing skills, students can practice skills to conduct online research which include forming a research question, locating online information, evaluating the information for selection, synthesizing the selected information, and communicating findings [54].

Another observation, relevant to the creative nature of synthesis in education, is the influential role knowledge synthesis plays in the development of the creative arts and humanities. For example, I propose that a composer synthesizes music out of musical components such as rhythm, timbre, pitch, and harmony. A painter synthesizes a painting out of form, texture, perspective, and color. A poet synthesizes a poem out of language, metaphor, and emotion. Artists often synthesize their style from the styles of the artistic giants who influenced them. Languages themselves are synthesized from other languages, social scientists such as psychologists and economists synthesize their work from the work of their predecessors (e.g., Freud and Marx), and so on. All fields in the arts and humanities have the potential to benefit from the knowledge synthesis methods described in this paper.

4. Synthesizing an Explanatory Theory

Psychologist Kurt Lewin proclaimed, "There is nothing so practical as a good theory" [55]. This perspective article's method of knowledge synthesis borrows heavily from theoretical research methods such as grounded theory. Glaser and Strauss developed the grounded theory method to bring more rigor to qualitative research [56], although the methodological principles of grounded theory are also applicable to quantitative research [57]. The researcher's overarching aim in grounded theory is to begin an investigation with a clean slate and inductively construct a new theory through an iterative process of comparative data analysis. With a new theory in hand, the researcher can encourage the development of hypotheses to experimentally test concepts deducted from the theory. Eventually, systematic reviews and meta-analyses can critically assess results of experiments and clinical trials testing the concepts. Interestingly, Ioannidis [58] suggested that the number of systematic reviews of clinical trials may be currently higher than the number of clinical trials, implying a greater need for theory synthesis at the source in the flow of new scientific information. Of relevance, the number of physician-scientists has also been on the decline over the past several decades, further highlighting the need to increase physician medical education in scientific thinking and biomedical research [59], including a need for education in knowledge synthesis and theory development.

Researchers most often use grounded theory as a method to analyze, compare, and combine concepts from original data, but Wolfswinkel et al. proposed that grounded theory may also be used to conduct a rigorous literature review in which concepts from published research findings themselves are the data [60]. Reviewing literature in this manner is particularly useful for synthesizing new knowledge from current knowledge—similar to thematic synthesis [36], a qualitative method that combines thematic analysis with meta-ethnography across multiple studies. An important difference is that the method proposed by Wolfswinkel et al. is a mixed method that combines both qualitative and quantitative research.

The following is an example of how I used a grounded theory approach to conduct a web-based synthesis in the field of epidemiology. I was interested in exploring online research literature investigating the association of phosphate toxicity with cancer [61]. The literature was reviewed using keyword searches in Google, Google Scholar, and other scholarly online databases, and all relevant studies associating phosphorus, tumorigenesis, cancer, etc. were identified. In addition to searching with keywords, I searched references cited in studies, which is known as citation analysis; a very useful method for evaluating a study's impact in a research area [62]. For example, The Hallmarks of Cancer [63] is an influential study often cited by other studies in a literature review of cancer research.

Included in my synthesis were studies selected from the fields of basic research, clinical research, and epidemiological research; the designs of the selected studies included case studies, cohort studies, laboratory animal experiments, in vitro studies, systematic reviews and meta-analyses. My investigation followed the evidence without boundaries, and my final synthesis was written in the form of a narrative review. Although many speculations, hypotheses, and explanatory theories were proposed by authors of the studies selected for my synthesis, I analyzed only concepts from the objective findings of the selected studies in order to foster a new grounded theory. The rigorous evidence-based grounded theory method helped assure the high internal validity of the synthesis, i.e., the analyzed concepts were based on trustworthy peer-reviewed findings rather than speculation. As concepts from the reviewed literature were analyzed and compared, certain themes began to emerge from the data which were eventually linked together into a cohesive theory that explained how phosphate toxicity from dysregulated phosphorus metabolism stimulated cancer cell growth. Interestingly, this inferred knowledge challenged several of the hypotheses proposed in The Hallmarks of Cancer. For example, evidence from my synthesis supported the concept that cancer cell growth is dependent on exogenous growth-rate factors, challenging the concept that cancer cells independently stimulate themselves to grow autonomously.

5. Domain-Specific Knowledge

To extend current knowledge into new knowledge using the synthesis method introduced in this perspective article, possessing a solid foundational base of expert knowledge in one's field, called domain-specific knowledge [64], is vital. Most narrative reviews are written by experts [65], and research in memory and learning has revealed some important clues about how expertise is acquired. Practicing repeated retrieval of memorized information over a period of time strengthens knowledge storage and recall from long-term memory, developing the kind of deep learning possessed by experts which enables them to comprehend complex concepts, solve difficult problems, and infer new associations [66]. This finding implies that knowledge synthesis can be strengthened by repeated readings and practiced memory recall of relevant selected literature that lies outside the scope of one's domain-specific knowledge. Nevertheless, opportunities to form novel, potentially ground-breaking transdisciplinary theories and theses appear remote unless one is willing to investigate research outside of one's domain of expertise during the process of new knowledge synthesis.

Logan [67] criticized the disconnect between specialized knowledge and meaningful context, quoting Konrad Lorenz:

"Philosophers are people who know less and less about more and more, until they know nothing about everything. Scientists are people who know more and more about less and less, until they know everything about nothing."

In other words, a proper balance of both knowledge depth and scope is necessary for meaningful context and new knowledge synthesis. Research confirms that boundary-spanning knowledge search methods, which expand beyond one's domain-specific range, promote the discovery of new pathways and new combinations of knowledge that lead to breakthroughs [68]. For example, based on my web-based synthesis of phosphate toxicity and cancer, my awareness of the efficacy of a reduced-phosphate diet to treat chronic kidney disease patients led to a boundary-spanning proposal to test a similar phosphate-reduced diet as a novel intervention for cancer prevention and treatment [69].

6. Association, Causation, and Mediation

An essential function of scientific inquiry is the ability to recognize patterns that connect discrete pieces of data and provide new meaning to the investigation [70]. Connecting pieces of evidence together in a synthesis is analogous to assembling pieces of a jigsaw puzzle to create a coherent big picture. The graphical abstract in Figure 2 uses the analogy of a jigsaw puzzle to illustrate how current, prior, qualitative, and quantitative knowledge are pieced together from various domains during knowledge synthesis. This section of the article provides an elementary explanation of relationships by which research concepts are logically combined during synthesis into new knowledge. The strengths and weaknesses of linked concepts can strengthen or weaken one's synthesis. I provide further examples from my web-based syntheses in the field of epidemiology, and the strengths and limitations of three common types of linked relationships are discussed: association, causation, and mediation.

6.1. Association

An association is a link between two items or variables. Sometimes the link may be coincidental or spurious and occur strictly by chance, and sometimes a link may be meaningful and occur with a higher probability than by chance alone. An example of a coincidental association is demonstrated by flipping a coin. The outcome of heads or tails is strictly a matter of chance, assuming you are using a fair coin. Even if turning up 99 heads in a row, the chance that the next coin flip will turn up tails does not increase; it still remains at approximately 50%. Betting that there is a meaningful association between the number of coin flips and the chances that heads or tails will turn up next is known as a gambler's fallacy [71].

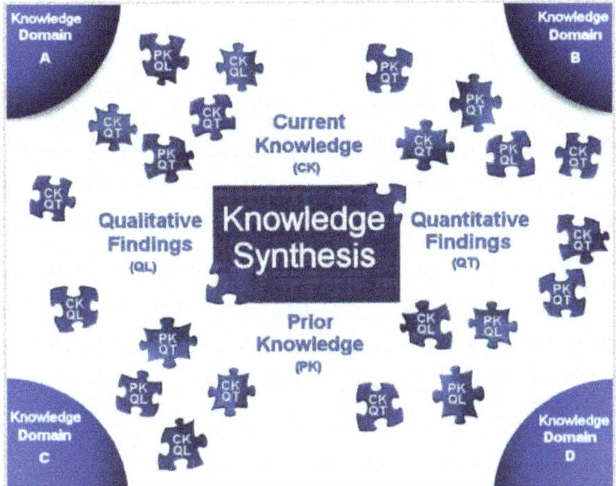

Figure 2. Graphical abstract. Like a jigsaw puzzle, pieces of knowledge are synthesized to form a coherent picture of new knowledge.

When associating concepts in a synthesis, the aim is to form a relationship in which the variables are meaningfully associated—as one variable changes, the associated variable also changes to some degree, known as covariance. But this does not necessarily mean that one variable is causing the other to change. Let us say the average person gets two colds a year and also takes vitamin C supplements. You do not take vitamin C supplements, and you observe that you get more colds than the average person. You suspect that taking vitamin C supplements may be associated with a reduced number of colds per year. But even if you are able to confirm that these two variables are statistically associated, as claimed in advertising using results of a clinical study, you still may not know what is causing the association between the variables. Perhaps people who take vitamin supplements look after their overall health better than you, which could be a confounding factor that independently causes the same outcome of reduced colds; which brings us to the next section.

6.2. Causation

Causation is a type of association in which one item, an independent variable, directly causes an effect on another item, a dependent variable or outcome variable. The highest form of evidence demonstrating causation is the randomized controlled trial (RCT), but RCTs are not always practical or feasible in research settings. Bradford Hill suggested the criteria in the list shown below for inferring causation from observational evidence in epidemiology [72]:

1. **Strength of the association**: stronger associations are more likely causal.
2. **Consistency**: causal findings are similar across multiple studies.
3. **Specificity of the association**: a cause produces a specific result.
4. **Temporality**: the cause precedes the result.
5. **Biological gradient**: the quantity of the cause determines the result size.
6. **Plausibility**: the cause may be explained by current knowledge.
7. **Coherence**: the cause is a good fit with related knowledge.
8. **Experimental evidence**: the cause occurs under controlled conditions.
9. **Analogy**: the cause is observed under similar circumstances.

6.3. Mediation

A mediator is a variable that lies between other variables within a direct causal pathway to an outcome variable. A directed acyclic graph (DAG) may be used to visually represent direct causal pathways between variables [73]. Acyclic means that a variable's causal pathway does not cycle back directly onto itself. Figure 3 is a DAG that shows a simple mediator causation pathway, based on Baron and Kenny, 1986 [74]. Note that confounders and effect modifiers lie outside the causation pathway in this model. For example, the independent variable may be replaced with an associated independent variable, a confounder, which causes the same effect on the outcome variable. An effect modifier may also change the outcome variable at the end of the causation pathway, as in the modifying effect of age and gender in the association of a risk factor with a disease. When inferring causation during a synthesis, possessing expert knowledge of the subject matter under investigation enables the identification of potential confounding factors [75], effect modifiers, and mediators. Research designs that include participant randomization and stratification of results can also assist in controlling the effects of confounders and effect modifiers.

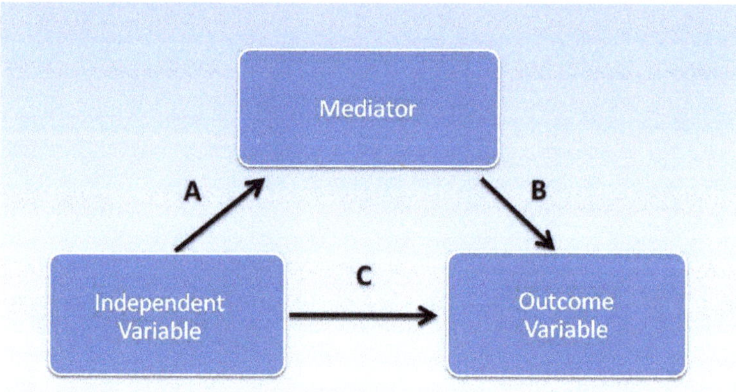

Figure 3. Mediation causal pathway. Causal path C runs from the independent variable to the outcome variable. Causal paths A and B run to and from the mediator, respectively, linking the independent and outcome variables. The absence of the mediator weakens path C, and if the path disappears altogether, the variables are linked indirectly through the mediator.

As causal diagrams have developed, indirect links between variables may be represented by a dotted line, and double-headed solid arrows may represent linked variables with an unspecified common cause [76]. I have also combined double-headed arrows with a dotted line (◄ − − − − − ►) to represent variables linked indirectly with an unspecified common cause. When conducting a synthesis, the researcher may infer the mediating common cause that indirectly links two variables. To illustrate, low vitamin D levels in patients have been associated with a higher risk of cancer incidence [77]. Based on this association, some researchers have proposed that taking vitamin D supplements may prevent cancer, but recently published clinical trials of vitamin D supplements and cancer prevention do not support this causal inference [78–80]. Having coauthored a textbook chapter on the endocrine regulation of phosphate homeostasis [81], I have background knowledge of vitamin D's role in regulating intestinal absorption of dietary phosphorus—i.e., vitamin D levels are lowered if phosphorus serum levels rise too high, as in clinical and subclinical hyperphosphatemia. Synthesizing the link between lowered vitamin D and hyperphosphatemia with the link between hyperphosphatemia and tumorigenesis [61], I proposed that hyperphosphatemia is a common cause that mediates an indirect association between lowered vitamin D levels and increased cancer risk [69].

When selecting information during knowledge synthesis, conflicting material helps identify areas requiring further in-depth investigation. As demonstrated in the above example of vitamin D supplementation and cancer prevention, there may be additional factors that are missing which thwart the synthesis of a truer overall picture. To illustrate, in the allegory of six blind-men and the elephant, each blind man examined a different part of the elephant by touch: the tail, trunk, tusk, ear, leg, and side, and each man inferred a different description of the nature of an elephant as being like a rope, snake, spear, fan, tree, and wall, respectively. Although their tactile observations were accurate, the men were unable to discover the true overall nature of an elephant because they did not synthesize their findings into new knowledge.

Mediation is also used in literature-based discovery, a synthesis method in which implicit knowledge is discovered from linking together separate bodies of literature [82]. For example, if concept A is related to concept B in one body of literature, and a separate body of literature relates the same concept B to concept C, transitive inference relates A to C, as shown in Figure 4. In this example, B acts as a potential mediator that causatively links two separate bodies of literature in a novel way to infer new knowledge.

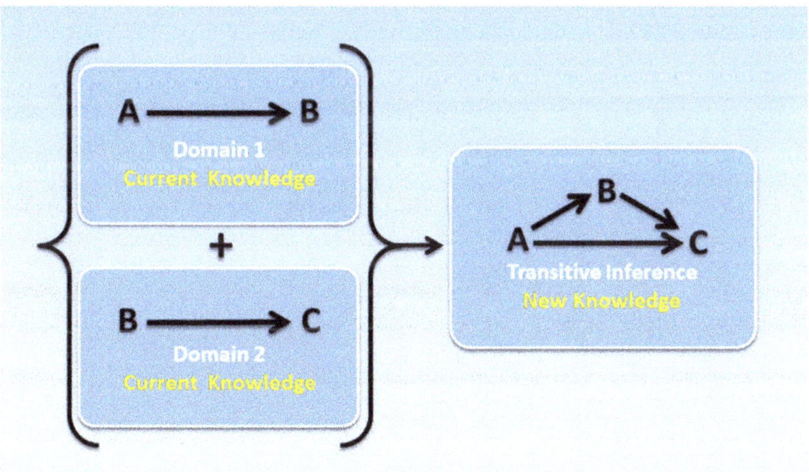

Figure 4. Transitive inference. A causes B in current knowledge domain 1, and B causes C in current knowledge domain 2. New knowledge is discovered when synthesizing domains 1 and 2 through transitive inference; i.e., A causes C, with B acting as a potential mediator.

I used transitive inference to propose an explanatory theory of how cholesterol oxidation products (COPs) are causatively linked to atherosclerosis [83]. I synthesized concepts from one body of research relating COPs (A) to defects in arterial cell membranes (B) with another body of research relating defects in arterial cell membranes (B) to atherosclerosis (C). In this case, defects in arterial cell membranes (B) acted as a mediator that linked COPs with atherosclerosis. This synthesis helped fill in some theoretical knowledge gaps in the potential cause and mechanism of atherosclerosis and strengthened the evidence for dietary prevention of atherosclerosis by avoiding COPs in thermally treated and processed animal-based foods that contain cholesterol.

7. Conclusions

Breakthrough knowledge has been shown to occur most often when prior knowledge is mixed with current knowledge. The art of discovering breakthrough knowledge in the age of Google involves writing web-based syntheses using interactive tools like online search engines and online databases.

Synthesis writing is an important educational objective. Grounded theory is a useful synthesis method to select, organize, analyze, and combine concepts from a mixture of research findings to infer an explanatory theory. Having domain-specific knowledge is important for synthesizing new knowledge and for identifying potential confounding factors, effect modifiers, and mediators when inferring a causal theory. I provided several examples of knowledge syntheses from my own published web-based syntheses in epidemiology, demonstrating how this novel method is readily available for use by researchers in other fields.

Funding: This research received no external funding.

Conflicts of Interest: The author declares no conflicts of interest.

References

1. Grimm, S.R.; Baumberger, C.; Ammon, S. *Explaining Understanding: New Perspectives from Epistemology and Philosophy of Science*; Taylor & Francis: Abingdon, UK, 2016.
2. Burnyeat, M. *The Theaetetus of Plato*; Hackett Publishing: Indianapolis, IN, USA, 1990.
3. Gettier, E.L. Is justified true belief knowledge? *Analysis* **1963**, *23*, 121–123. [CrossRef]
4. Turnbull, H.W. *The Correspondence of Isaac Newton: 1661–1675*; Cambridge University Press: Cambridge, UK, 1959.
5. Mukherjee, S.; Romero, D.M.; Jones, B.; Uzzi, B. The nearly universal link between the age of past knowledge and tomorrow's breakthroughs in science and technology: The hotspot. *Sci. Adv.* **2017**, *3*, e1601315. [CrossRef] [PubMed]
6. Markos, L. God and Galileo: What a 400-Year-Old Letter Teaches Us about Faith and Science. *Christ. Sch. Rev.* **2019**, *49*, 99–102.
7. Kuhn, T.S. *The Structure of Scientific Revolutions*; University of Chicago Press: Chicago, IL, USA, 1962.
8. Dewar, J.A. *The Information Age and the Printing Press: Looking Backward to See Ahead*; RAND Corporation: Santa Monica, CA, USA, 1998.
9. Shrivastav, H.; Hiltz, S.R. Information overload in technology-based education: A meta-analysis. In Proceedings of the 19th Americas Conference on Information Systems, Chicago, IL, USA, 15–17 August 2013.
10. Williams, R. *Keywords (Routledge Revivals): A Vocabulary of Culture and Society*; Routledge: London, UK, 2013.
11. Garson, D.; Lillvik, C.; Sisk, E.; Ewing, E.; Johnson, M. The Literature Review: A Research Journey. Harvard Graduate School of Education. 2016. Available online: https://guides.library.harvard.edu/literaturereview (accessed on 27 April 2019).
12. Kastner, M.; Tricco, A.C.; Soobiah, C.; Lillie, E.; Perrier, L.; Horsley, T.; Welch, V.; Cogo, E.; Antony, J.; Straus, S.E. What is the most appropriate knowledge synthesis method to conduct a review? Protocol for a scoping review. *BMC Med. Res. Methodol.* **2012**, *12*, 114. [CrossRef]
13. Sutton, A.J.; Abrams, K.R. Bayesian methods in meta-analysis and evidence synthesis. *Stat. Methods Med. Res.* **2001**, *10*, 277–303. [CrossRef]
14. Stemler, S. An overview of content analysis. *Pract. Assess. Res. Eval.* **2001**, *7*, 137–146.
15. Dixon-Woods, M.; Cavers, D.; Agarwal, S.; Annandale, E.; Arthur, A.; Harvey, J.; Hsu, R.; Katbamna, S.; Olsen, R.; Smith, L. Conducting a critical interpretive synthesis of the literature on access to healthcare by vulnerable groups. *BMC Med. Res. Methodol.* **2006**, *6*, 35. [CrossRef]
16. Droitcour, J.; Silberman, G.; Chelimsky, E. Cross-design synthesis: A new form of meta-analysis for combining results from randomized clinical trials and medical-practice databases. *Int. J. Technol. Assess. Health Care* **1993**, *9*, 440–449. [CrossRef]
17. Banning, J.H. Ecological triangulation: An approach for qualitative meta-synthesis. Available online: http://citeseerx.ist.psu.edu/viewdoc/download?doi=10.1.1.152.5185&rep=rep1&type=pdf (accessed on 4 February 2020).
18. Pope, C.; Ziebland, S.; Mays, N. Qualitative research in health care: Analysing qualitative data. *BMJ Br. Med. J.* **2000**, *320*, 114. [CrossRef]
19. Strauss, A.; Corbin, J. *Basics of Qualitative Research Techniques: Techniques and Procedures for Developing Grounded Theory*; Sage publications: Thousand Oaks, CA, USA, 1998.

20. Noblit, G.W.; Hare, R.D. *Meta-Ethnography: Synthesizing Qualitative Studies*; Sage: Thousand Oaks, CA, USA, 1988; Volume 11.
21. Weed, M. "Meta Interpretation": A Method for the Interpretive Synthesis of Qualitative Research. *Forum Qual. Soc. Res.* **2005**, *6*, 1–21.
22. Greenhalgh, T.; Robert, G.; Macfarlane, F.; Bate, P.; Kyriakidou, O.; Peacock, R. Storylines of research in diffusion of innovation: A meta-narrative approach to systematic review. *Soc. Sci. Med.* **2005**, *61*, 417–430. [CrossRef] [PubMed]
23. Paterson, B.L.; Thorne, S.E.; Canam, C.; Jillings, C. *Meta-Study of Qualitative Health Research: A Practical Guide to Meta-Analysis and Meta-Synthesis*; Sage: Thousand Oaks, CA, USA, 2001; Volume 3.
24. Sandelowski, M.; Barroso, J. Creating metasummaries of qualitative findings. *Nurs. Res.* **2003**, *52*, 226–233. [CrossRef] [PubMed]
25. Sandelowski, M.; Docherty, S.; Emden, C. Qualitative metasynthesis: Issues and techniques. *Res. Nurs. Health* **1997**, *20*, 365–371. [CrossRef]
26. Pluye, P.; Gagnon, M.-P.; Griffiths, F.; Johnson-Lafleur, J. A scoring system for appraising mixed methods research, and concomitantly appraising qualitative, quantitative and mixed methods primary studies in mixed studies reviews. *Int. J. Nurs. Stud.* **2009**, *46*, 529–546. [CrossRef] [PubMed]
27. Dixon-Woods, M.; Agarwal, S.; Jones, D.; Young, B.; Sutton, A. Synthesising qualitative and quantitative evidence: A review of possible methods. *J. Health Serv. Res. Policy* **2005**, *10*, 45–53. [CrossRef]
28. Popay, J.; Roberts, H.; Sowden, A.; Petticrew, M.; Arai, L.; Rodgers, M.; Britten, N.; Roen, K.; Duffy, S. Guidance on the conduct of narrative synthesis in systematic reviews. In *Results of An ESRC Funded Research Project*; Unpublished Report; University of Lancaster: Lancaster, UK, 2006.
29. Yin, R. *Case Study Research–Design and Methods*; Applied Social Research Methods Series; Sage Publications: Thousand Oaks, CA, USA, 2003; Volume 5.
30. Jensen, L.A.; Allen, M.N. Meta-synthesis of qualitative findings. *Qual. Health Res.* **1996**, *6*, 553–560. [CrossRef]
31. Grant, M.J.; Booth, A. A typology of reviews: An analysis of 14 review types and associated methodologies. *Health Inf. Libr. J.* **2009**, *26*, 91–108. [CrossRef]
32. Yin, R.K.; Heald, K.A. Using the case survey method to analyze policy studies. *Adm. Sci. Q.* **1975**, *20*, 371–381. [CrossRef]
33. Pawson, R.; Greenhalgh, T.; Harvey, G.; Walshe, K. Realist review-a new method of systematic review designed for complex policy interventions. *J. Health Serv. Res. Policy* **2005**, *10*, 21–34. [CrossRef]
34. Lucas, P.J.; Baird, J.; Arai, L.; Law, C.; Roberts, H.M. Worked examples of alternative methods for the synthesis of qualitative and quantitative research in systematic reviews. *BMC Med. Res. Methodol.* **2007**, *7*, 4. [CrossRef]
35. Mays, N.; Pope, C.; Popay, J. Systematically reviewing qualitative and quantitative evidence to inform management and policy-making in the health field. *J. Health Serv. Res. Policy* **2005**, *10*, 6–20. [CrossRef] [PubMed]
36. Thomas, J.; Harden, A. Methods for the thematic synthesis of qualitative research in systematic reviews. *BMC Med. Res. Methodol.* **2008**, *8*, 45. [CrossRef] [PubMed]
37. Wyborn, C.; Louder, E.; Harrison, J.; Montambault, J.; Montana, J.; Ryan, M.; Bednarek, A.; Nesshöver, C.; Pullin, A.; Reed, M. Understanding the Impacts of Research Synthesis. *Environ. Sci. Policy* **2018**, *86*, 72–84. [CrossRef]
38. Schick-Makaroff, K.; MacDonald, M.; Plummer, M.; Burgess, J.; Neander, W. What synthesis methodology should I use? A review and analysis of approaches to research synthesis. *AIMS Public Health* **2016**, *3*, 172. [CrossRef] [PubMed]
39. Downs, F.S.; Fawcett, J. *The Relationship of Theory and Research*; McGraw-Hill/Appleton & Lange: London, UK, 1986.
40. Hjørland, B. Domain analysis. *Ko Knowl. Organ.* **2017**, *44*, 436–464. [CrossRef]
41. Martínez, V.; Berzal, F.; Cubero, J.-C. A survey of link prediction in complex networks. *ACM Comput. Surv. (CSUR)* **2016**, *49*, 1–33. [CrossRef]
42. Liben-Nowell, D.; Kleinberg, J. The link-prediction problem for social networks. *J. Am. Soc. Inf. Sci. Technol.* **2007**, *58*, 1019–1031. [CrossRef]
43. Qi, Y.; Bar-Joseph, Z.; Klein-Seetharaman, J. Evaluation of different biological data and computational classification methods for use in protein interaction prediction. *Proteins Struct. Funct. Bioinform.* **2006**, *63*, 490–500. [CrossRef] [PubMed]

44. Cumming, A.; Lai, C.; Cho, H. Students' writing from sources for academic purposes: A synthesis of recent research. *J. Engl. Acad. Purp.* **2016**, *23*, 47–58. [CrossRef]

45. Bloom, B.; Krathwohl, D. *Taxonomy of Educational Objectives: The Classification of Educational Goals by A Committee of College and University Examiners*; Domain, H.I.C., Ed.; Longman: New York, NY, USA, 1956.

46. Forehand, M. Bloom's taxonomy. *Emerg. Perspect. Learn. Teach. Technol.* **2010**, *41*, 47–56.

47. Carpenter, S.R.; Armbrust, E.V.; Arzberger, P.W.; Chapin, F.S., III; Elser, J.J.; Hackett, E.J.; Ives, A.R.; Kareiva, P.M.; Leibold, M.A.; Lundberg, P. Accelerate synthesis in ecology and environmental sciences. *Bioscience* **2009**, *59*, 699–701. [CrossRef]

48. Zee, M.; De Boer, M.; Jaarsma, A. Acquiring evidence-based medicine and research skills in the undergraduate medical curriculum: Three different didactical formats compared. *Perspect. Med Educ.* **2014**, *3*, 357–370. [CrossRef]

49. National Commission on Writing in America's Schools and Colleges. Writing: A ticket to work … or a ticket out: A survey of business leaders. 2004. Available online: http://www.writingcommission.org/prod_downloads/writingcom/writing-ticket-to-work.pdf (accessed on 7 November 2019).

50. National Commission on Writing. Writing: A Powerful Message from State Government. Available online: http://www.writingcommission.org/prod_downloads/writingcom/powerful-message-from-state (accessed on 7 November 2019).

51. Magrath, C.; Ackerman, A.; Branch, T.; Clinton Bristow, J.; Shade, L.; Elliott, J.; Williams, R. *The Neglected "R": The Need for A Writing Revolution*; College Entrance Examination Board; The National Commission on Writing: New York, NY, USA, 2003.

52. Luo, L.; Kiewra, K.A. *A SOAR-Fired Method for Teaching Synthesis Writing*; IDEA Paper# 74; IDEA Center, Inc.: Fort Recovery, OH, USA, 2019.

53. Lai, C.C.; Lik, H.C.; Pheng, O.S.; Shee, L.S.; Joyce, T.T.; Rosli, N. Enhancing classroom engagement through web-based interactive tools. *Proc. Mech. Eng. Res. Day 2019* **2019**, *2019*, 218–220.

54. Leu, D.J.; Kinzer, C.K.; Coiro, J.; Castek, J.; Henry, L.A. New literacies: A dual-level theory of the changing nature of literacy, instruction, and assessment. *J. Educ.* **2017**, *197*, 1–18. [CrossRef]

55. Marrow, A.J. *The Practical Theorist: The Life and Work of Kurt Lewin*; Teachers College Press: New York, NY, USA, 1977.

56. Glaser, B.G.; Strauss, A.L. *Discovery of Grounded Theory: Strategies for Qualitative Research*; Routledge: New York, NY, USA, 2017. [CrossRef]

57. Chun Tie, Y.; Birks, M.; Francis, K. Grounded theory research: A design framework for novice researchers. *SAGE Open Med.* **2019**, *7*. [CrossRef] [PubMed]

58. Ioannidis, J.P. The mass production of redundant, misleading, and conflicted systematic reviews and meta-analyses. *Milbank Q.* **2016**, *94*, 485–514. [CrossRef]

59. Schafer, A.I.; Mann, D.L. Current Education of Physicians: Lost in Translation? *JACC Basic Transl. Sci.* **2019**, *4*, 655–657. [CrossRef]

60. Wolfswinkel, J.F.; Furtmueller, E.; Wilderom, C.P. Using grounded theory as a method for rigorously reviewing literature. *Eur. J. Inf. Syst.* **2013**, *22*, 45–55. [CrossRef]

61. Brown, R.B.; Razzaque, M.S. Phosphate toxicity and tumorigenesis. *Biochim. Biophys. Acta Rev. Cancer* **2018**. [CrossRef]

62. Bu, Y.; Waltman, L.; Huang, Y. A multidimensional perspective on the citation impact of scientific publications. *arXiv* **2019**, arXiv:1901.09663.

63. Hanahan, D.; Weinberg, R.A. The hallmarks of cancer. *Cell* **2000**, *100*, 57–70. [CrossRef]

64. Ackerman, P.L. Domain-specific knowledge as the "dark matter" of adult intelligence: Gf/Gc, personality and interest correlates. *J. Gerontol. Ser. B Psychol. Sci. Soc. Sci.* **2000**, *55*, 69–84. [CrossRef] [PubMed]

65. Green, B.N.; Johnson, C.D.; Adams, A. Writing narrative literature reviews for peer-reviewed journals: Secrets of the trade. *J. Chiropr. Med.* **2006**, *5*, 101–117. [CrossRef]

66. Karpicke, J. A powerful way to improve learning and memory: Practicing retrieval enhances long-term, meaningful learning. *Psychol. Sci. Agenda.* Available online: http://www.apa.org/science/about/psa/2016/06/learning-memory.aspx (accessed on 9 December 2019).

67. Logan, R.K. In Praise of and a Critique of Nicholas Maxwell's In Praise of Natural Philosophy: A Revolution for Thought and Life. *Philosophies* **2018**, *3*, 20. [CrossRef]

68. Katila, R.; Ahuja, G. Something old, something new: A longitudinal study of search behavior and new product introduction. *Acad. Manage. J.* **2002**, *45*, 1183–1194.
69. Brown, R.B. Vitamin D, cancer, and dysregulated phosphate metabolism. *Endocrine* **2019**, 1–6. [CrossRef]
70. Harwood, W.S.; Reiff, R.R.; Phillipson, T. Putting the Puzzle Together: Scientists' Metaphors for Scientific Inquiry. *Sci. Educ.* **2005**, *14*, 25–30.
71. Lepley, W.M. "The Maturity of the Chances": A Gambler's Fallacy. *J. Psychol.* **1963**, *56*, 69–72. [CrossRef]
72. Fedak, K.M.; Bernal, A.; Capshaw, Z.A.; Gross, S. Applying the Bradford Hill criteria in the 21st century: How data integration has changed causal inference in molecular epidemiology. *Emerg. Themes Epidemiol.* **2015**, *12*, 14. [CrossRef]
73. Rohrer, J.M. Thinking clearly about correlations and causation: Graphical causal models for observational data. *Adv. Methods Pract. Psychol. Sci.* **2018**, *1*, 27–42. [CrossRef]
74. Baron, R.M.; Kenny, D.A. The moderator–mediator variable distinction in social psychological research: Conceptual, strategic, and statistical considerations. *J. Pers. Soc. Psychol.* **1986**, *51*, 1173. [CrossRef] [PubMed]
75. Robins, J.M. Data, design, and background knowledge in etiologic inference. *Epidemiology* **2001**, *12*, 313–320. [CrossRef] [PubMed]
76. Greenland, S.; Pearl JBoslaugh, S. Causal diagrams. In *Encyclopedia of Epidemiology*; Boslaugh, S.E., Ed.; Calif SAGE Publications: Thousand Oaks, CA, USA, 2008.
77. Bikle, D.D. Vitamin D and cancer: The promise not yet fulfilled. *Endocrine* **2014**, *46*, 29–38. [CrossRef] [PubMed]
78. Manson, J.E.; Cook, N.R.; Lee, I.-M.; Christen, W.; Bassuk, S.S.; Mora, S.; Gibson, H.; Gordon, D.; Copeland, T.; D'Agostino, D. Vitamin D supplements and prevention of cancer and cardiovascular disease. *N. Engl. J. Med.* **2019**, *380*, 33–44. [CrossRef]
79. Scragg, R.; Khaw, K.-T.; Toop, L.; Sluyter, J.; Lawes, C.M.; Waayer, D.; Giovannucci, E.; Camargo, C.A. Monthly high-dose vitamin D supplementation and cancer risk: A post hoc analysis of the vitamin D assessment randomized clinical trial. *JAMA Oncol.* **2018**, *4*, e182178. [CrossRef]
80. Keum, N.; Lee, D.; Greenwood, D.; Manson, J.; Giovannucci, E. Vitamin D supplementation and total cancer incidence and mortality: A meta-analysis of randomized controlled trials. *Ann. Oncol.* **2019**. [CrossRef]
81. Brown, R.B.; Razzaque, M.S. Chapter 31—Endocrine Regulation of Phosphate Homeostasis. In *Textbook of Nephro-Endocrinology*, 2nd ed.; Singh, A.K., Williams, G.H., Eds.; Academic Press: Cambridge, MA, USA, 2018; pp. 539–548. [CrossRef]
82. Swanson, D.R. Undiscovered public knowledge. *Libr. Q.* **1986**, *56*, 103–118. [CrossRef]
83. Brown, R.B. Phospholipid packing defects and oxysterols in atherosclerosis: Dietary prevention and the French paradox. *Biochimie* **2019**. [CrossRef]

 *philosophies*

Article

A Cognitive Perspective on Knowledge How: Why Intellectualism Is Neuro-Psychologically Implausible

Andreas Stephens [1,*] **and Cathrine V. Felix** [1,2,*]

[1]   Department of Philosophy, Lund University, 22100 Lund, Sweden
[2]   Department of Philosophy and Religious Studies, Norwegian University of Science and Technology, 2815 Gjøvik, Norway
*   Correspondence: andreas.stephens@fil.lu.se (A.S.); cathrine.felix@ntnu.no (C.V.F.)

Received: 25 July 2020; Accepted: 24 August 2020; Published: 5 September 2020

Abstract: We defend two theses: (1) Knowledge how and knowledge that are two distinct forms of knowledge, and; (2) Stanley-style intellectualism is neuro-psychologically implausible. Our naturalistic argument for the distinction between knowledge how and knowledge that is based on a consideration of the nature of slips and basic activities. We further argue that Stanley's brand of intellectualism has certain ontological consequences that go against modern cognitive neuroscience and psychology. We tie up our line of thought by showing that input from cognitive neuroscience and psychology, on multiple levels of analysis, cohere in supporting the distinction between two separate forms of knowledge. The upshot is a neuro-psychologically plausible understanding of knowledge.

Keywords: intellectualism; anti-intellectualism; knowledge; knowledge how; knowledge that; naturalism; slips; basic activities

---

1. Introduction

Knowledge is commonly seen as consisting of two forms. The knowledge how to do something, such as riding a bike, and the knowledge that something is the case, such as that Reykjavík is the capital of Iceland.[1] Ryle [1,2] set an enduring and influential modern debate in motion by discussing the relation between the two forms. According to intellectualism, which Ryle opposes, knowledge how is a form of knowledge that. According to anti-intellectualism on the other hand, which Ryle supports, knowledge how involves abilities and dispositions, and is thereby not a form of knowledge that (see, e.g., [1–3]):

> Cleverness at fighting is exhibited in the giving and parrying of blows, not in the acceptance or rejection of propositions about blows … [n]or does the surgeon's skill function in his tongue uttering medical truths but only in his hands making the correct movements. [2] (p. 48).

Stanley and Williamson [3–5] offers a thought-provoking intellectualist account of knowledge, where they—against Ryle—argue that knowing-how is a form of propositional knowing-that. We will focus on Stanley's [3] formulation, where he presents an intellectualist account arguing that facts are true propositions and that knowledge of how to do things is knowledge of facts:

> [K]nowing how to do something is the same as knowing a fact. It follows that learning how to do something is learning a fact. For example, when you learned how to swim, what happened

---

[1]   The distinction is standard in the literature and we will use it throughout the article. However, there are, of course, other conceptualizations of knowledge including, for example, knowledge by *acquaintance*, as well as knowing *which, who, what, what* it is like etc.

is that you learned some facts about swimming. Knowledge of these facts is what gave you knowledge of how to swim. Something similar occurred with every other activity that you now know how to do, such as riding a bicycle or cooking a meal. You know how to perform activities solely in virtue of your knowledge of facts about those activities. [3] (p. vii).

As Schwartz and Drayson [6] point out, it is crucial to make one's own stance in the intellectualist/anti-intellectualist debate clear since a lot of confusion arguably stems from theoreticians talking past each other depending on whether their focus is on semantics or science. In philosophy, numerous different approaches for investigating knowledge have been discussed and promoted, however, the most common focus has been on the *concept* knowledge, and Stanley's argument centers on a semantic approach. This said, he explicitly discusses cognitive scientific evidence and claims that it supports his position [3,7]. Therefore, this approach is motivated to assess Stanley's claim using a naturalistic approach where the *natural phenomenon* knowledge, such as science presents it, is seen as being of primary interest.[2] Our approach is thus cooperative, and complementarily evolutionary, naturalistic in the sense that we consider it essential to take empirical evidence and scientific theories into account—trumping intuitions—when such input is available in order to find a plausible account of knowledge. All philosophical questions are then legitimate, although their answer will need to be compatible with scientific input [8–14].

There have been a number of naturalistic attacks on Stanley's position (see, e.g., [15–17]), yet while we largely agree with most of these arguments, we want to add a pluralistic perspective that highlights consideration of the nature of slips and basic activities as well as the scientific multi-level coherence for the case against intellectualism.

In short, we claim that the intellectualist position of knowledge how is untenable in light of relevant scientific theorizing and evidence. In defense of our view, we put forth and defend the following two theses: (1) knowledge how and knowledge that should be seen as two distinct forms of knowledge, and (2) intellectualism is neuroscientifically and psychologically implausible.

2. Stanley's Intellectualism

We will follow Stanley [3] in focusing on procedural knowledge how and propositional knowledge that, although both phenomena can be, and have been, conceptualized along, for example, the following lines concerning knowledge and memory: tacit/implicit versus explicit; non-declarative/procedural versus declarative; practical versus theoretical. These various pairs involve subtle differences in connotation. However, it is often hard to pinpoint the exact usage in debates where the various concepts tend to be used interchangeably. However, given our naturalistic approach, this is somewhat less problematic since it, in our view, is ultimately the natural world that governs how the natural phenomenon knowledge should be understood.

Stanley claims that a reluctance to accept the intellectualist view often depends on an incorrect understanding of what knowledge of facts amounts to, i.e., not properly accounting for the connection between knowledge of facts and action. It is, according to Stanley, special kinds of facts—propositions answering the relevant "how-questions"—that enables knowledge how. Stanley points out that:

> Humans are thinkers and humans are agents. There is a natural temptation to view these as distinct capacities, governed by distinct cognitive states. When we engage in reflection, we

---

[2] While most naturalists acknowledge the natural world as being "all-encompassing," and science as providing our best method of understanding it, there are a number of different interpretations of where this leaves philosophy [8]. An *ontological* naturalism that rejects anything supernatural is commonly accepted although there is still debate whether, for example, the mental can be reduced to the physical [9]. However, the proper form of *methodological* naturalism is a much debated issue, involving positions which argue for *replacing* epistemology with cognitive psychology [10], *substantially* reformulating philosophical problems into scientific terminology [11], or for *cooperatively* taking scientific input into account whenever such information is available [12]. Moreover, *evolutionary* naturalism has highlighted how cognitive agents are shaped by natural selection—information which might *replace, succeed* or *complement* epistemology [13,14].

are guided by our knowledge of propositions. By contrast, when we engage in intelligent action, we are guided by our knowledge of how to perform various actions. If these are distinct cognitive capacities, then knowing how to perform an action is not a species of propositional knowledge. [3] (p. 1, italics in original).

As we will show below, it is actually the case that engaging in reflection and intelligent action does involve distinctly different cognitive capacities, which means that intellectualism will be hard to defend.

Stanley positions himself against Ryle [1,2], but what he formulates is first and foremost an empirical question. Therefore, we will look to cognitive psychology and cognitive neuroscience as the most relevant sources of input and answers. Stanley does discuss how cognitive scientific evidence ought to inform philosophy, but his interpretation is problematic for reasons we will discuss below. In short, we claim that a general problem with Stanley's approach is that it conflates the possibility of describing an agent performing an action with an agent actually being *able* to perform that action (see, e.g., [17], Section 4).

Stanley does qualify his position and wants to separate knowledge how from involving ability, and argues that "knowing how to do something does not entail being able to do it" [3] (p. 127). This is, however, not a satisfactory position. Indeed, it is correct that, for example, a pianist, in a sense, might know how to play the piano even after losing their arms. However, contrary to Stanley, this would be due to neural pathways that have been entrained by numerous hours spent behind the keys, not due to propositional facts that the pianist "thinks of practically". They might be able to describe how someone else ought to go about it to play a certain tune, but unless that person, in turn, actually put in the work to get similar relevant neural pathways that person would be unable to actually play, and therefore not know how to play (see, e.g., [18]). In addition, since Stanley [3] (p. 131) explicitly claims that his "view of the nature of knowing how to do something is a view about the metaphysical nature of these states, and not a view in semantics" we believe that it involves several problematic features.

Stanley's focus on language and questioning of the relevance of scientific input is problematic from a naturalistic perspective like the one we defend. He discusses the matter thus:

Why would one expect it to be a virtue of an account of knowing how that it is plausibly taken to be what is expressed by ascriptions of knowing how in natural language? Shouldn't we be open to the possibility that science could show us that states of knowing how are very different in kind from what ordinary speakers use sentences like "Ana knows how to swim" to express? If so, then although it might be a virtue of an account of the meaning of ascriptions of knowing how that it is plausibly correct for natural language, it is not a virtue of an account of the nature of states of knowing how. And it is the latter that concerns the philosopher, and not the former. In various forms and versions, this is the foundational objection to the methodology I have employed. [3] (pp. 143–144).

Stanley [3] (p. 144, italics in original) concludes that "[i]t could hardly be that science could discover that knowing how to swim was a distinct state than is expressed by 'knowing how to swim'. After all, quite minimal principles governing the nature of the truth-predicate render this incoherent."

However, this is non-naturalistic argumentation that seems to rely completely on a certain set of intuitions as a foundation. To let the intuitive usage of language dictate—and overrule— in advance, what science conceives of as a natural phenomenon is a radical position that Stanley does not present convincing support for. Importantly, he at other places wants to use cognitive scientific input in defense of his position, which we will get back to below (see, e.g., [3] (ch. 7), [7]). His initial commitment to view facts as propositions seems to have led him astray. It is possibly this position that renders Stanley unaccommodating towards Noë's [16] critique of his (and Williamson's) focus on language, ascriptions of knowledge, and a semantic understanding of propositionality (see also, e.g., [19]), rather than the nature of the mind and knowledge seen as a natural phenomenon [3] (p. 146). It seems to be this interpretation, where cognitive science only can amount to different understandings of "relations between propositions" [3] (p. 148), that makes Stanley evade the conclusion from the scientific theories he himself quotes ([20] (p. 209), [21] (pp. 562–563), see also [6,16,17]); knowing how and knowing that are actually different phenomena that involve different cognitive processes, and the two forms of knowledge are explicitly held apart by science (see, e.g., [20,22–25], see also [16,26]).

Stanley [3] (p. 150) claims that "knowing how to do something is a kind of propositional knowledge, a kind of propositional knowledge that guides skilled actions." He argues that this is the case by claiming that declarative knowledge is but one form of propositional knowledge; procedural (non-declarative) knowledge can also be propositional, or so Stanley argues. In order to make this account fit better with scientific evidence, he goes on to highlight that not all propositional knowledge needs to be capable of being verbalized [3] and that knowing how to do something is conceptual. It is *prima facie* reasonable to assume that an intellectualist of Stanley's disposition would defend the view that knowledge how includes an ability to verbalize the knowledge in question (see also footnote 3). However, it is sufficient on Stanley's account that the agent in question can make use of demonstrative expressions like "this", exemplified by Stanley thorough the imagined case of the young Mozart who, when asked how he is composing his music merely points to one of his masterpieces and says "*this* is how I can do it" (emphasise is ours). In another paper ("A Naturalistic Perspective on Knowledge How: Grasping Truths in a Practical Way" [28]), we argue that one would expect from an intellectualist account like Stanley's, that the agent can express their knowledge how in a more illuminating sense than by the mere use of demonstrative expressions. Stanley, however, simply states that if one accepts that all words count, then demonstrative expressions should be accepted as ways of expressing one's knowledge how [4] (p. 214). In fact, Stanley argues that words are not needed *at all* to demonstrate knowledge how. He tries to show this through the example of an expert boxer who fights a southpaw. By boxing the exact way he does, the expert proves through his agency that he knows how to fight a southpaw. He does not need to say how he knows or what he does when fighting, because he is, as a matter of fact doing it, due to skill, not mere luck. As we write in our paper, we find this approach unsatisfactory given the overall intellectualist framework Stanley defends.

With some caution, we can concede that not all propositional knowledge needs to be capable of being verbalized. Importantly, though, this hinges on exactly how one interprets a number of factors, such as, for example, how knowledge is connected to various forms of long-term memory, and their correlating neurological substrate, among other issues. For example, depending on how one draws the line concerning propositional knowledge and its connection to episodic and/or semantic long-term memory, different interpretations come out as being most plausible.

Stanley introduces an interesting point when he writes that the distinction between procedural and propositional knowledge does not involve two forms of states but rather two ways of generating or implementing a particular state of knowledge [3] (p. 151). This is a possible interpretation, but should ultimately be evaluated with a focus on the actual underlying cognitive phenomena (processes). By merely focusing on very high-level aspects of knowledge Stanley seems to be drawn to biologically implausible conclusions. This is, for example, salient when he in a footnote writes that "[b]eing merely procedurally implemented cannot be sufficient for knowing how" [3] (p. 153, fn. 2, italics in original). Simple mechanical processes, on this account, are then not enough for knowledge how. However, this point is debatable. Many organisms, for example, have simple reflexive mechanisms that enable them to interact with their environment in a way that, arguably, ought to be viewed as involving knowing how to do something—a point we will get back to below (see, e.g., [29,30]).

---

3 Stanley [4] (p. 213) quotes Fodor [27] (p. 634): "There is a real and important distinction between knowing how to do a thing and knowing how to explain to do that thing. But *that* distinction is one that the intellectualist is perfectly able to honor . . . The ability to give explanation is itself a skill—a special kind of knowing how which presupposes general verbal facility at the very least. But what has this to do with the relation between knowing how and knowing *that*? And what is there here to distress an intellectualist?".

The view we defend can be developed further to include animal cognition, for example along the lines developed by Jan Faye's evolutionary naturalist approach [30]. However, we will not discuss the topic in detail in this paper. It is a project for another paper. Some details regarding Stanley's account should be mentioned though. In a brief passage in his book *Know How* [3] (p. 133), Stanley discusses Alva Noë's critique that Stanley and Williamson [5], given their intellectualism, are forced to say that animals can grasp propositions because animals obviously knows how to do many things, a dog, for example, knows how to catch a frisbee. The point is not that animals cannot grasp propositions, but rather that this is a highly controversial issue. Stanley's reply seems to dodge the critique. His reply is, in short, that automaticity of action and animal cognition fits all varieties of knowing-wh (where, when, whether etc.). Stanley remarks that a philosophical account of knowledge how that says that it is not a kind of propositional knowledge will have the unlucky result that many cases of knowing-wh too do not ascribe propositional knowledge [3] (p 134). Does Stanley bite the bullet, or not? Given what he says, it is hard to tell.

3. Basic Activities and Slips

Basic activities are of special interest to our enquiry. Basic activities are those activities we automatically know how to perform; we pull them off by doing nothing else. They are the very source of agency, the material of which all other actions are made. As such they are a necessary condition for all other matters connected to agency. When someone performs basic activities, they do so by making bodily movements that constitute the fundamental atoms of what they are attempting to accomplish, grounding every other action. They do so without any need to determine the means by which the action should be performed. Basic activities are such that the agent knows how to perform them, say, "just like that." Consider the example from Hornsby of someone saying "grass is green" [31]. The utterance is performed through movements of the tongue, mouth, teeth, etc. The agent does not make these movements because they believe that they are a means of saying "grass is green"; they probably have no particular beliefs about them and are hardly aware of them. They do not do them intentionally in the way they intentionally utters the sentence as a whole. From the agent's perspective, basic activities have a teleological structure, and one can perform them without any need to understand their details correctly. On such a teleological view, basic activities are relative to the agent: acts that are basic for one agent may not be for another. Moreover, something that is basic for an agent in one setting may not be in another. Consider the agent for whom using their own hot-water kettle is a basic action while using a hot-water kettle of unfamiliar design is not.

The possibility of slips is of special interest to the naturalistic view we defend. Slips—performances one can successfully do while intending something else—genuinely reveal what one can do in a basic sense, "just like that", precisely because one has all the necessary bodily skills to perform them. Because of that capacity, one can reasonably be expected to be able to perform them intentionally as well. A basic activity is one the agent can perform even while making a slip. We think those things that one is able to do even while making a slip are the ones that truly are basic. In other words, "slipping behavior" reveals basicness. A slip is the kind of performance that would often, in another context, be meaningfully intended. After all, slips are not random. Verbal slips show lexical biases, tend to result in grammatically correct utterances, and rarely violate the syntactic constraints of the language; so Fromkin [32] (p. 183) writes: "according to all linguists who have analyzed spontaneous speech errors, the errors are nonrandom and predictable". This is important, because it illuminates how an agent's knowledge how is operative when they slip. When an agent slips by grasping the salt instead of the sugar, then this is not because they are incompetent in distinguishing salt from sugar. What the slip shows is misdirected competence. Slips are small episodes of misdirected behavior in otherwise correctly performed action sequences, often closely resembling the correct act. Salt is not so far from sugar in terms of appearance or use. Contrast the act of grasping the salt instead of the sugar with that of accidentally taking a handful of soil from a pot of tulips: such action surely would be baffling, because it differs so much from what is expected. The point becomes particularly clear with so-called slips of the tongue: the agent who slips speaks comprehensible words, not mere nonsense.

To specify even more, slips are executive failures; you go left instead of right, you pour milk instead of water, you push the button marked "1" instead of the one marked "2" etc. In the case of a slip, the agent bluntly does something different from what they mean to be doing. The failure is not at a cognitive level, thus the agent does not fail because they have a false belief and the like [33–36]. Crucially, the agent slips despite knowing that their performance is an error. The agent who is about to add milk to their coffee, but pours juice instead does not err because they cannot tell the difference between milk and juice: clearly, they can. Competence does not provide immunity from slips: the person who slips does not do so because they lack the relevant ability. They are fully able to grab, lift, and pour from a container; they just happen to grab the wrong one. They slip not because their movements are beyond their control, like those made by someone suffering from a compulsive disorder or Anarchic Hand syndrome,[4] but because, then and there, they simply make the wrong movements. Among other things, an important thing to notice here is that the agent has the correct propositional content and is normally able to apply it, but, for some reason, their knowledge that and their knowledge how came apart and they acted in some sense successfully: they performed actions that belong to their basic action repertoire (they pull them off "just like that")—i.e., they poured milk instead of juice, they did not start singing into the jug or jump up and down—on a different propositional content than they meant to act on. How can knowledge that and -how come apart like this if knowledge how is merely a species of the former? The topic of basicness in general and the above discussed split between knowledge how and knowledge that in particular is not discussed by Stanley. He does not consider the basicness of activities the way we do. Instead of focusing on basicness, he focuses on knowledge of truths and is ignorant towards executive mistakes in the sense we are interested in here, it just is not among his concerns. He states that when you learn some truths about swimming, then you know how to swim and after the acquisition of that fact it seems, on his intellectualist account, that you can, of course, do something wrong when swimming, boxing and cycling, but, crucially, you cannot bluntly apply the wrong content and be wrong about how you should swim or cycle. When you have learned how to swim, you simply know. In other words, when you have gained the knowledge of swimming, you cannot "unknow" it, in the same way as after the moment you genuinely learned the truths about how to walk, you cannot be wrong about walking, to walk is something you know how to do. One of our points is that Stanley's account can make one lose sight of neuro-psychological facts like the simple point that human beings sometimes slip and in the slip, knowledge how and knowledge that come apart, as we have shown in the above examples. The failure to see this is one of the things that we find biologically misconceived in the intellectualist account. This is, in a nutshell, our critique: slips show how knowledge how has epistemic properties not present in knowledge that. When an agent slips, they do something different from what they intended; nonetheless, the performance is guided by their knowledge how. This reveals a divide between the knowledge that actively guides behavior: the knowledge how that the agent applies sub-consciously; and the knowledge how they intend to guide their behavior in the first place, which they are under the illusion of acting on even as they slips. We argue that this divide between two levels of knowledge how operative in the slip case has no parallel when it comes to knowledge that. Therefore, knowledge how cannot be reduced to knowledge that. As we will explain in further detail below, a reason for why slips occur can be a slip-up in the communication between different cognitive processes in the mind of the agent at the time of acting.

---

[4] Addressing Anarchic Hand syndrome, Marcel [37] (p. 77) writes: "the affected hand performs unintended but complex, well-executed, goal-directed actions. Often when the patient is trying to do something with the unaffected hand, the other hand appears to do the opposite or compete with it."

4. Grounding Knowledge

By looking to cognitive psychology and cognitive neuroscience we can gain a lot of insight regarding the cognitive underpinnings of knowledge (see, e.g., [38–45]). In particular, we will argue that various influential theories focusing on different levels of analysis provide a multi-level coherent picture. From a first high-level perspective, dual process theory offers elucidating input for our discussion. According to this influential theory,[5] cognitive processes consist in Type 1 processes (System 1) and Type 2 processes (System 2). Type 1 processes are fast, reflexive, and non-conscious, whereas Type 2 processes are slow, reflective, and conscious. These different processes are plausibly mapped to different forms of knowledge. In short, knowledge how can plausibly be tentatively mapped to reflexive Type 1 processes and knowledge that can plausibly be tentatively mapped to reflective Type 2 processes (see, e.g., [46,47], see also [48]):

> Motor behavior, moreover, can take place without the experience of intentional states or reasoning, not only in the case of true reflexes but also, importantly, in the case of expert performance, such as that of an athlete "in the zone" or a chess master. [49] (p. 34)

On a lower level of analysis, the human memory systems follow the same pattern (see, e.g., [20,43–45]. Tulving's account states that long-term memory consists in three parts working together—procedural, semantic, and episodic memory—where " … procedural memory entails semantic memory as a specialized subcategory, and … semantic memory, in turn, entails episodic memory as a specialized subcategory." [43] (pp. 2–3). Type 1 processes can plausibly be mapped to procedural memory involving bottom-up action schemas, abilities, and skills that are possible to learn through trial and error. Type 2 processes can plausibly be mapped to episodic memory involving propositional, factual remembrance, and the experienced first-person point of view. As an intertwined intermediate, semantic memory involves conceptual content and categorization. Using Cohen and Squire's ([20], see also [22]) canonical conceptualization, this division can be classified as non-declarative (or procedural) versus declarative memory:

> Declarative memory includes what can be declared or brought to mind as a proposition or an image. … Non-declarative memory refers to a heterogeneous collection of abilities: motor skills, perceptual skills, and cognitive skills (these abilities and perhaps others are examples of procedural memory); as well as simple classical conditioning, adaptation/eye/effects, pdming, and other instances where experience alters performance independently of providing a basis for the conscious recollection of past events. [22] (p. 171)

On yet a lower level of analysis, procedural memory involves, for example, the basal ganglia, neocortex, cerebellum, striatum, and the premotor- and primary motor cortex (see, e.g., [50]). Semantic memory involves, for example, associative pathways, the prefrontal cortex, the lateral-, ventral- and medial temporal cortex, basal ganglia, and hippocampus (see, e.g., [50]). Episodic memory involves, for example, attentional pathways, the prefrontal, ventral fronto-temporal, medial temporal, retrosplenial, and posterior cingulate cortices, the parahippocampal, angular, middle temporal, the fusiform, and inferior temporal gyrus, as well as the left posterior insula and the hippocampus (see, e.g., [50]).

To use motor output as an example, thoughts, intentions, and knowledge that involve Type 2 processes, episodic memory, the frontal lobe and cortex, and varying layers of smaller and larger homotypical cells. Associations and conceptual knowledge involve semantic memory, the parietal lobe and association cortex, and varying layers of smaller and larger homotypical cells. Whereas knowledge how involves Type 1 processes, procedural memory, the primary motor cortex, cerebellum, and large

---

[5]   We are aware of critical voices (see, e.g., [38,42], see also the discussion below).

agranular cells (see, e.g., [50]). Investigations on multiple levels of analysis, and from different scientific perspectives, thus support a division of knowledge how and knowledge that.

Now, this understanding of memory and knowledge has recently been criticized by De Brigard [51]—although his focus is specifically on the non-declarative (procedural)/declarative view of the memory systems, which he calls the standard model of memory (SMM). We consider many of his points interesting and meriting further investigation that might ultimately lead to a new interpretation of the memory systems. However, note that the SMM account differs in certain details from Tulving's account which we have presented above. We therefore believe that De Brigard's critique loses a lot of its bite when this is taken into consideration. Regarding the intellectualist position specifically, we do consider De Brigard's arguments to be underwhelming since he presents a very dogmatic interpretation of what he dubs the "empirical argument." On his interpretation, this line of reasoning demands that *"P2. Knowledge how is equivalent to (or, at least, exclusively depends on) procedural [non-declarative] memory, whereas knowledge that is equivalent to (or, at least, exclusively depends on) declarative memory."* [51] (p. 728). We believe this premise to be a misleading interpretation and would change the phrasings "equivalence" and "exclusive dependence" to something less strong and domineering. For the anti-intellectualist argument, formulations in the line of that "knowledge how is primarily based in (and primarily depends on) procedural memory" would suffice. This is enough to counter the intellectualist position. Importantly, we also question De Brigard's [51] (p. 728) third premise *"P3. But the scientific evidence captured by the SMM demonstrates that procedural (non-declarative) and declarative memory are two entirely dissociable and independent systems."* As we have briefly described above, Tulving's account of memory posits that long-term memory involves procedural, semantic, and episodic memory, and that the three systems work in parallel. However, importantly, procedural memory is evolutionarily prior to semantic and episodic memory which means that if any reduction is to be made, it ought to go the complete opposite way from how intellectualists see things (see, e.g., [52,53]).

Stanley [3] argues that because episodic and semantic memory are separate forms of declarative memory and knowledge, not all declarative knowledge is propositional (only semantic memory and knowledge is). This is a reasonable, albeit not undisputed, interpretation. However, it does not follow that procedural memory and knowledge should be viewed as being propositional. By gaining propositional knowledge it is possible to gain in skill, but only after using this propositional knowledge to actually perform actions repeatedly. Stanley's (see, e.g., [3], pp. 155–159) characterization of this process vastly downplays the reflexive and unconscious processes involved in acquiring procedural knowledge and vastly exaggerates the role of propositional knowledge in skillful actions. Stanley argues for his interpretation, where (explicit) procedural knowledge should be viewed as propositional, but does not present empirical evidence supporting this interpretation. He rather bases it on his discussions of how people ascribe knowledge how to others. However, empirical evidence instead indicates the non-propositional nature of procedural memory and knowledge, where "[d]eclarative or explicit knowledge is available to consciousness, and can readily be expressed verbally, whereas procedural or implicit knowledge is typically unconscious and can only be expressed verbally with some difficulty and sometimes not at all" ([23], p. 65; see, e.g., [20,23–25], see also, e.g., [6,16,17,26]). Devitt [16] sums up the situation as follows:

> Despite disagreement or uncertainty on many other issues, psychologists speak with one voice on this one. Even production-system theories, which seem to posit representations of processing rules and hence seem to have the most "intellectualized" picture of procedural knowledge, distinguish this knowledge sharply from declarative knowledge. Psychology presents a picture of procedural knowledge as constituted somehow or other by embodied, probably unrepresented rules that are inaccessible to consciousness. It is thus quite different from declarative knowledge which consists of representations that are available to consciousness. ... In sum, a nonpropositional view of knowledge how is not just

philosophical prejudice or even just folk theory: it seems to be entrenched in psychology and cognitive ethology. [16] (pp. 213–215)

In other words, procedural (non-declarative) processes do not involve brain regions and processes linked to language in the same way as declarative memory and knowledge do. Stanley's semantic reading and idiosyncratic stipulation of what procedural and propositional knowledge involves thus runs against empirical evidence and scientific theorizing, and mischaracterizes the number of non-conscious processes involved in procedural knowledge how (see, e.g., [6] (Section 4.2), [15] (p. 287), [17] (Section 2)).

Devitt [16] argues for claims that are similar to those we have presented, pointing out that it is problematic to base one's view of knowledge solely on linguistic arguments[6] concerning ascriptions of knowledge. Devitt [16] (p. 206) further argues that knowledge how does not reduce to knowledge that, but does temper his claim to state that " ... perhaps one kind of knowledge how is knowledge that. But I still want to maintain that another kind is not. There is a common kind of knowledge how that a person can have simply on the basis of having the ability to perform an activity." It would then be wrong to claim that all knowledge how reduces to knowledge that.

We are sympathetic to Devitt's view, but maybe he gives too much slack to the intellectualist position? After all, all knowledge how involves vast amounts of non-conscious processes (see, e.g., [24,55]:

> There is now considerable evidence to suggest that the performance of reward-related actions in both rats and humans reflects the interaction of two quite different learning processes, one controlling the acquisition of goal-directed actions, and the other the acquisition of habits. This evidence suggests that, in the goal-directed case, action selection is governed by an association between the response 'representation' and the 'representation' of the outcome engendered by those actions, whereas in the case of habit learning, action selection is controlled through learned stimulus–response (S–R) associations without any associative link to the outcome of those actions. As such, actions under goal-directed control are performed with regard to their consequences, whereas those under habitual control are more reflexive in nature, by virtue of their control by antecedent stimuli rather than their consequences. [56] (p. 49)

The intellectualist thus ignores empirical evidence and scientific theorizing, and is thus, at its core, a problematic methodology (see, e.g., [57]).

It is clear that knowledge how for Stanley is substantially tied to language use. Even though it is not a requirement that the agent in question is able to express in words their knowledge how, it is required that they, in some sense, are both a language-user and have a concept of self. This is so because having knowledge how for Stanley amounts to knowing the answer to a question, and moreover, to apply the acquired knowledge to oneself (realizing that I can Φ). In sum, it seems that Stanley must hold that the concepts of language and self are necessarily involved in having knowledge how. This, by comparison, is not a demand on our view since knowledge how on our account basically rests on procedural memory (see Section 4). The claim that knowing how to do something is conceptual is, or so we argue, potentially problematic or even false. We do believe that the bulk of (cognitive) scientific input strongly indicates that animals are capable of conceptualizing their environment, while not being capable of language, and that Stanley's position is deeply problematic. As we point out, regardless of how one draws the line concerning declarative memory (semantic and/or episodic LTM), it does not

---

[6] Stanley and Williamson ([5], p. 441) write that their account follows "from basic facts about the syntax and semantics of ascriptions of knowledge how". It springs from a theory of embedded questions: according to which knowing how is treated as on a par with knowing whether, knowing when and so on. We do not offer an extensive analysis of Stanley's linguistic arguments here. Our view, in short, sides with Glick's [54] view, saying that using mere linguistic premises, about knowledge how ascriptions, to support substantive conclusions about the nature of knowledge how just cannot do the job it is supposed to do.

follow that non-declarative memory (procedural LTM), which is most reasonably tied to knowledge how, or so we argue in the paper, is conceptional, and even less so; propositional. In sum, we hold the view that verbalization is not necessary for conceptualization. Moreover, we hold that there are plausible arguments to be had for both positions, but that we want to leave the question open to further debate. Importantly, however, we argue that it is up to Stanley to prove that non-propositional knowledge how is reasonable to tie to propositional knowledge that—which we argue that he does not succeed with. In other words, it strikes us that the burden of proof is on Stanley, where we consider his account to be idiosyncratic and not in line with a naturalistic interpretation of relevant scientific input.

5. Grounding Slips

Recall the phenomenon of slips. The cognitive processes underlying knowledge how and knowledge that, and how they come apart, can be elucidated by this phenomenon:

> The 'slips of action' that punctuate daily life illustrate that our behaviour is not always goal-directed in nature. Folk wisdom suggests that such slips of action occur when well-practised responses intrude to compromise our goal-directed behaviour. For example, it is a common-place experience to find oneself arriving at the door of the old office although one's original intention was to get to the new one. Adams [58] was the first to show that in rats extensive training of the instrumental response of lever pressing rendered it impervious to devaluation of the food outcome, a finding that has now been replicated in a number of rodent studies. [59] (p. 471).

Goal-directed control of actions is important to ensure that one's behavior has the desired result. However, experiments by de Wit et al. [60] show that such control competes with habitual responses. In dual process terminology; reflective Type 2 processes can be "hijacked" by reflexive Type 1 processes, which results in slips of action. Importantly, de Wit et al. point out that a reason for such slips can be that although Type 2 processes allow for greater flexibility, Type 1 processes demand less cognitive effort. Loosely speaking, the system (agent) might thus automatically fall back into default mode in order to save energy. If this is done incorrectly the agent slips. Empirical studies[7] suggest that dopamine plays a central role in regulating the interplay between the cognitive processes regarding this matter, and how slips occur (see, e.g., [60,61]).[8]

Rasmussen [62] (see also [48], p. 43) offers a three-level model that classifies human behavior as being either skill-based, rule-based, or knowledge-based. Skill-based behavior consists in non-conscious automatic actions demanding extensive training. Rule-based behavior, on an intermediate level, consists in actions guided by information acquired empirically or through communicated instructions. Knowledge-based behavior consists in goal-controlled actions that do not rely on know how or rules but rather are based on reflective cognitive processes.

Focusing on knowledge how, Rasmussen's account can be interpreted as presenting an olive branch towards Stanley's [3] and Stanley and Williamson's [5] accounts in that knowledge how might plausibly be seen as consisting in both a procedural element (procedural memory) and a rule-based element (semantic memory):

> In general, the skill-based performance rolls along without the person's conscious attention, and he will be unable to describe how he controls and on what information he bases the performance. The higher level rule-based coordination is generally based on explicit know how, and rules used can be reported by the person. [62] (p. 259)

Where Stanley and Williamson go wrong is that they ignore the procedural element and only focus on the rule-based element. Moreover, they ascribe features more suitably linked to knowledge-based

---

[7] Studies have primarily been done on animals, but those that focus on humans indicate that generalizability is plausible.
[8] This said, for example, serotonin is also thought to play an important role.

behavior to knowledge how, and finally claim that it is possible to reduce knowledge how to knowledge that which is completely unsubstantiated.

Representing actions with schemas, Norman (1981)[9] develops a similar account:

> A major assumption of the ATS [activation-trigger-schema] theory for slips is that skilled actions—actions whose components are themselves all highly skilled—need only be specified at the highest levels of their memory representations. Once the highest-level schema is activated, the lower-level parent components of that action sequence complete the action, to a large extent autonomously, without further need for intervention except at critical choice points. [36] (p. 4)

Mismatches from episodic memory (false beliefs) can, arguably, be seen as involving knowledge that. This is so since formations of intentions involve reflective features that we have argued above is linked to episodic long-term memory and working memory. These types of mismatches can usually be fixed by focused attention and additional cognitive energy expenditure. Slips due to mismatches from procedural memory are more readily seen to involve knowledge how, where a particular action schema is executed–but not the intended one. In these cases, we are thus in possession of relevant knowledge how, but we accidentally carry out some other knowledge how (schema) that we also possess.

6. The Neuro-Psychological Implausibility of Intellectualism

Humans are but one animal among others. It is therefore essential to take animal cognition research into account when investigating knowledge how and knowledge that. As we have argued above, knowledge can be seen to be grounded in the memory systems. An important point regarding this topic is that procedural memory (linked to knowledge how) is evolutionarily prior to semantic and episodic memory (linked to knowledge that).

Accordingly, many animals are capable of knowledge how without having a language, and thus arguably not having knowledge that. Once again, if there is a reduction to be made, it should therefore reduce knowledge that to knowledge how rather than the other way around (see, e.g., [52,53]).

Now, Stanley and Krakauer [7] try to wedge knowledge how apart by focusing on skilled actions, where they argue that motor skill demands propositional knowledge that of facts whereas motor acuity ("precision of selected actions") does not. Contrary to canonical cognitive scientific interpretations keeping propositional knowledge and skill apart, Stanley and Krakauer [7] (p. 1) argue that " … learning to become skilled at a motor task, for example tennis, depends also on knowledge-based selection of the right actions[,]" and then go on to argue that " … skilled activity requires both acuity and knowledge, with both increasing with practice." We concede that Stanley and Krakauer have a point in that knowledge that indeed might increase acuity, but that it only can do so to a comparatively very limited—indeed infinitesimally small—amount. It is therefore, in our view, misleading to push this detail in the manner Stanley and Krakauer do. Stanley and Krakauer even go further and promote an idiosyncratic view, that appears contrary to well-established interpretations in cognitive psychology and cognitive neuroscience, in claiming that "the identification of procedural knowledge, itself a misnomer, with the colloquial understanding of motor skill … [and] the identification of declarative knowledge with knowledge in the traditional sense [are incorrect]" [7] (p. 2):

> Does the fact that manifesting skill requires knowledge preclude non-human animals from possessing skills? We are agnostic as to whether animals can be skilled. It is possible that as a task is weighted increasingly toward rules, alternative actions, and on-the-fly problem

---

[9] According to Norman [36], slips in the form of intentional formational errors involve cases where a specific action schema is intended but vagueness and contextual factors intervene. Slips, in the form of faulty activations, involves actions that result from unintended schemas that might be similar to or associated with the intended schema, or are due to the loss of activation. Slips, in the form of faulty triggerings, can result both from failures to trigger schemas or by triggering at inappropriate circumstances.

solving, then simple operant conditioning may not suffice to accomplish the task. In this sense non-human animals may be limited in a way similar to the amnestic patients in the Roy and Park experiment. Although non-human animals may exhibit the same behavior as humans, this does not entail that the explanation for the behavior is the same. It could be that the explanation skilled action in humans involves intellectual capacities lacking in non-human animals. ... Alternatively, it could be that animals can both possess concepts and bear the knowledge relation to propositions (if so, one would need to explain why animals cannot acquire certain skills that humans can; perhaps because there complex skills require complex concepts, which cannot be grasped by animals). [7] (p. 9)

As Noë highlights, abilities are embodied and situated in ways that are deeply problematic for intellectualism [15] (p. 284), and contrary to the intellectualist position, we have shown that procedural knowledge how is a distinct form of knowledge from propositional knowledge that, involving separate brain areas, cognitive processes, and cell-types (see, e.g., [56]), and these findings are generalizable across species:

> Rodent studies have implicated prelimbic cortex and its striatal efferents on dorsomedial striatum as a key circuit responsible for goal-directed learning. In a series of fMRI studies, vmPFC has been found to be involved in encoding reward predictions based on goal-directed action–outcome associations in humans, suggesting that this region of cortex in the primate prefrontal cortex is a likely functional homolog of prelimbic cortex in the rat. Furthermore, the area of anterior caudate nucleus found in humans to be modulated by contingency would seem to be a candidate homolog for the region of dorsomedial striatum implicated in goal-directed control in the rat. Finally, the evidence reviewed here supports the suggestion that a region of dorsolateral striatum in rodents and of the putamen in humans is involved in the habitual control of behavior, which when taken together with the findings on goal-directed learning reviewed previously, provides converging evidence that the neural substrates of these two systems for behavioral control are relatively conserved across mammalian species. [56] (p. 54).

7. Concluding Remarks

The gist of the present article has been to make a case for a more plausible understanding of knowledge. An understanding of knowledge that, to some extent, harmonizes with agents' subjective experience of their actions when it comes to such phenomena as slips and basic actions, but primarily in regards to cognitive psychological and neuroscientific findings on the nature of agency. This has led us to argue for a distinction between knowledge how and knowledge that, and against a Stanley-styled intellectualism. Our argument against the latter has been that Stanley's position has certain ontological consequences that do not equate well with neuroscience and psychology. Moreover, cognitive sciences support a distinction between knowledge how and knowledge that; a distinction that especially comes to the fore, we have argued, in the phenomenon of slips.

Author Contributions: Both authors have contributed to: Conceptualization, Methodology and Writing—Original draft preparation, Review and Editing. All authors have read and agreed to the published version of the manuscript.

Funding: A.S. gratefully acknowledges support from Makarna Ingeniör Lars Henrik Fornanders fond and Stiftelsen Elisabeth Rausings minnesfond: forskning.

Acknowledgments: We thank our anonymous reviewers for valuable comments.

Conflicts of Interest: The authors declare no conflict of interest.

References

1. Ryle, G. Knowing how and knowing that. *Proc. Aristot. Soc.* **1946**, *46*, 1–16. [CrossRef]
2. Ryle, G. *The Concept of Mind*; University of Chicago Press: Chicago, IL, USA, 1949.

3.   Stanley, J. *Know How*; Oxford University Press: Oxford, UK, 2011.

4.   Stanley, J. Knowing (how). *Noûs* **2011**, *45*, 207–238. [CrossRef]

5.   Stanley, J.; Williamson, T. Knowing how. *J. Philos.* **2001**, *98*, 411–444. [CrossRef]

6.   Schwartz, A.; Drayson, Z. Intellectualism and the argument from cognitive science. *Philos. Psychol.* **2019**, *32*, 661–691. [CrossRef]

7.   Stanley, J.; Krakauer, J.W. Motor skill depends on knowledge of facts. *Front. Hum. Neurosci.* **2013**, *7*, 1–11. [CrossRef] [PubMed]

8.   Rysiew, P. Naturalism in epistemology. In *The Stanford Encyclopedia of Philosophy*, 2017th ed.; Zalta, E.N., Ed.; CSLI: Stanford, CA, USA, 2017; Available online: https://plato.stanford.edu/archives/spr2017/entries/epistemology-naturalized (accessed on 25 August 2020).

9.   Stoljar, D. Physicalism. In *The Stanford Encyclopedia of Philosophy*, 2017th ed.; Zalta, E.N., Ed.; CSLI: Stanford, CA, USA, 2017; Available online: https://plato.stanford.edu/archives/win2017/entries/physicalism/ (accessed on 25 August 2020).

10. Quine, W.V.O. Epistemology naturalized. In *Ontological Relativity and Other Essays*; Columbia University Press: New York, NY, USA, 1969; pp. 69–90.

11. Churchland, P.S. Mind-brain reduction: New light from the philosophy of science. *Neuroscience* **1982**, *7*, 1041–1047. [CrossRef]

12. Kornblith, H. *Knowledge and its Place in Nature*; Oxford University Press: Oxford, UK, 2002.

13. Campbell, D.T. Evolutionary epistemology. In *The Philosophy of Karl R. Popper*; Schilpp, P.A., Ed.; Open Court: LaSalle, IL, USA, 1974; pp. 412–463.

14. Bradie, M.; Harms, W. Evolutionary epistemology. In *The Stanford Encyclopedia of Philosophy*, 2017th ed.; Zalta, E.N., Ed.; CSLI: Stanford, CA, USA, 2017; Available online: https://plato.stanford.edu/archives/spr2017/entries/epistemology-evolutionary (accessed on 25 August 2020).

15. Noë, A. Against intellectualism. *Analysis* **2005**, *65*, 278–290. [CrossRef]

16. Devitt, M. Methodology and the nature of knowing how. *J. Philos.* **2011**, *108*, 205–218. [CrossRef]

17. Roth, M.; Cummins, R. Intellectualism as cognitive science. In *Knowledge and Representation*; Newen, A., Bartels, A., Jung, E.-M., Eds.; CSLI: Stanford, CA, USA, 2011; pp. 23–39.

18. Fridland, E. Knowing-how: Problems and considerations. *Eur. J. Philos.* **2015**, *23*, 703–727. [CrossRef]

19. Glauer, R. The semantic reading of propositionality and its relation to cognitive-representational explanations: A commentary on Andreas Bartels and Mark May. In *Open MIND: 2(C)*; Metzinger, T., Windt, J.M., Eds.; MIND Group: Frankfurt am Main, Germany, 2015.

20. Cohen, N.; Squire, L. Preserved learning and retention of pattern-analyzing skill in amnesia: Dissociation of knowing how and knowing that. *Science* **1980**, *210*, 207–210. [CrossRef]

21. Tardif, T.; Wellman, H.M.; Fung, K.Y.F.; Liu, D.; Fang, F. Preschoolers' understanding of knowing-that and knowing-how in the United States and Hong Kong. *Dev. Psychol.* **2005**, *41*, 562–573. [CrossRef] [PubMed]

22. Squire, L.R.; Zola-Morgan, S. Memory: Brain systems and behavior. *Trends Neurosci.* **1988**, *11*, 170–175. [CrossRef]

23. Annett, J. On knowing how to do things: A theory of motor imagery. *Cogn. Brain Res.* **1995**, *3*, 65–69. [CrossRef]

24. Aarts, H.; Dijksterhuis, A. Habits as knowledge structures: Automaticity in goal-directed behavior. *J. Personal. Soc. Psychol.* **2000**, *78*, 53–63. [CrossRef]

25. Anderson, J.R. *Cognitive Psychology and its Implications*, 7th ed.; Worth Publishers: New York, NY, USA, 2009.

26. Wallis, C. Consciousness, context, and know how. *Synthese* **2008**, *160*, 123–153. [CrossRef]

27. Fodor, J. The appeal to tacit knowledge in psychological explanation. *J. Philos.* **1968**, *65*, 627–640. [CrossRef]

28. Felix, C.V.; Stephens, A. A Naturalistic Perspective on Knowledge How: Grasping Truths in a Practical Way. *Philosophies* **2020**, *5*, 5. [CrossRef]

29. Shettleworth, S.J. *Cognition, Evolution, and Behavior*; Oxford University Press: Oxford, UK, 2010.

30. Faye, J. *The Nature of Scientific Thinking. On Interpretation, Explanation and Understanding*; Palgrave Macmillan: London, UK, 2014.

31. Hornsby, J. *Actions*; Routledge & Kegan Paul: London, UK, 1980.

32. Fromkin, V. Slips of the Tongue. *Sci. Am.* **1973**, 181–187. Available online: http://web.stanford.edu/~{}zwicky/slips-of-the-tongue.pdf (accessed on 24 June 2015). [CrossRef]

33. Amaya, S. Slips. *Nôus* **2013**, *47*, 559–576. [CrossRef]

34. Anscombe, G.E.M. *Intention*; Harvard University Press: Harvard, IL, USA, 2000.
35. Baars, B.J. The many uses of error: Twelve steps to a unified framework. In *Experimental Slips and Human Error: Exploring the Architecture of Volition*; Baars, B.J., Ed.; Plenum Press: New York, NY, USA, 1992; pp. 3–38.
36. Norman, D.A. Categorization of action slips. *Psychol. Rev.* **1981**, *88*, 1–15. [CrossRef]
37. Marcel, A. The sense of agency: Awareness and ownership of action. In *Agency and Self-Awareness: Issues in Philosophy and Psychology*; Roessler, J., Eilan, N., Eds.; Oxford University Press: Oxford, UK, 2003; pp. 48–93.
38. Gigerenzer, G.; Regier, T. How do we tell an association from a rule? Comment on Sloman. *Psychol. Bull.* **1996**, *119*, 23–26. [CrossRef]
39. Keren, G.; Schul, Y. Two is not always better than one: A critical evaluation of two-system theories. *Perspect. Psychol. Sci.* **2009**, *4*, 533–550. [CrossRef] [PubMed]
40. Kruglanski, A.W.; Chun, W.Y.; Erb, H.P.; Pierro, A.; Mannett, L.; Spiegel, S. A parametric unimodel of human judgment: Integrating dual-process frameworks in social cognition from a single-mode perspective. In *Social Judgments: Implicit and Explicit Processes*; Forgas, J.P., Williams, K.R., von Hippel, W., Eds.; Cambridge University Press: New York, NY, USA, 2003; pp. 137–161.
41. Kruglanski, A.W.; Gigerenzer, G. Intuitive and deliberative judgements are based on common principles. *Psychol. Rev.* **2011**, *118*, 97–109. [CrossRef]
42. Barrett, H.C. *The Shape of Thought: How Mental Adaptations Evolve*; Oxford University Press: Oxford, UK, 2015.
43. Tulving, E. Memory and consciousness. *Can. Psychol.* **1985**, *26*, 1–12. [CrossRef]
44. Tulving, E. Episodic memory: From mind to brain. *Annu. Rev. Psychol.* **2002**, *53*, 1–25. [CrossRef] [PubMed]
45. Baddeley, A.D. *Working Memory, Thought and Action*; Oxford University Press: Oxford, UK, 2007.
46. Evans, J.S.B.T.; Stanovich, K.E. Dual-process theories of higher cognition: Advancing the debate. *Perspect. Psychol. Sci.* **2013**, *8*, 223–241. [CrossRef] [PubMed]
47. Lizardo, O.; Mowry, R.; Sepulvado, B.; Stoltz, D.S.; Taylor, M.A.; Van Ness, J.; Wood, M. What are dual process models? Implications for cultural analysis in sociology. *Sociol. Theory* **2016**, *34*, 287–310. [CrossRef]
48. Reason, J. *Human Error*; Cambridge University Press: Cambridge, UK, 1990.
49. Horst, S. *Cognitive Pluralism*; MIT Press: Cambridge, MA, USA, 2016.
50. Kandel, E.R.; Schwartz, J.H.; Jessell, T.M.; Siegelbaum, S.A.; Hudspeth, A.J. (Eds.) *Principles of Neural Science*, 5th ed.; McGraw-Hill, Health Professions Division: New York, NY, USA, 2013.
51. De Brigard, F. Know how, intellectualism, and memory systems. *Philos. Psychol.* **2019**, *32*, 720–759. [CrossRef]
52. Hetherington, S. Knowing-that, knowing-how, and knowing philosophically. *Grazer Philos. Stud.* **2008**, *77*, 307–324. [CrossRef]
53. Hetherington, S. *How to Know: A Practicalist Conception of Knowledge*; John Wiley & Sons: West Sussex, UK, 2011.
54. Glick, E. Two methodologies for evaluating intellectualism. *Philos. Phenomenol. Res.* **2011**, *83*, 398–434. [CrossRef]
55. Bargh, J.A. Goal ≠ intent: Goal-directed thought and behavior are often unintentional. *Psychol. Inq.* **1990**, *1*, 248–251. [CrossRef]
56. Balleine, B.W.; O'Doherty, J.P. Human and rodent homologies in action control: Corticostriatal determinants of goal-directed and habitual action. *Neuropsychopharmacology* **2010**, *35*, 48–69. [CrossRef] [PubMed]
57. Cath, Y. Knowing how. *Analysis* **2019**, *79*, 487–503. [CrossRef]
58. Adams, C.D. Variations in the sensitivity of instrumental responding to reinforcer devaluation. *Q. J. Exp. Psychol. Sect. B* **1982**, *34B*, 77–98. [CrossRef]
59. De Wit, S.; Dickinson, A. Associative theories of goal-directed behaviour: A case for animal–human translational models. *Psychol. Res. PRPF* **2009**, *73*, 463–476. [CrossRef] [PubMed]
60. De Wit, S.; Standing, H.R.; DeVito, E.E.; Robinson, O.J.; Ridderinkhof, K.R.; Robbins, T.W.; Sahakian, B.J. Reliance on habits at the expense of goal-directed control following dopamine precursor depletion. *Psychopharmacology* **2012**, *219*, 621–631. [CrossRef]
61. Nelson, A.; Killcross, S. Amphetamine exposure enhances habit formation. *J. Neurosci.* **2006**, *26*, 3805–3812. [CrossRef]
62. Rasmussen, J. Skills, rules, and knowledge: Signals, signs, and symbols, and other distinctions in human performance models. *IEEE Trans. Syst. Man Cybern.* **1983**, *13*, 257–266. [CrossRef]

 *philosophies*

Article

A Naturalistic Perspective on Knowledge How: Grasping Truths in a Practical Way

Cathrine V. Felix [1,2] and Andreas Stephens [2,*

[1] Department of Human Rights, Lund University, 221 00 Lund, Sweden; cathrine_v.felix@mrs.lu.se
[2] Department of Philosophy, Lund University, 221 00 Lund, Sweden
* Correspondence: andreas.stephens@fil.lu.se

Received: 18 February 2020; Accepted: 9 March 2020; Published: 12 March 2020

Abstract: For quite some time, cognitive science has offered philosophy an opportunity to address central problems with an arsenal of relevant theories and empirical data. However, even among those naturalistically inclined, it has been hard to find a universally accepted way to do so. In this article, we offer a case study of how cognitive-science input can elucidate an epistemological issue that has caused extensive debate. We explore Jason Stanley's idea of the *practical grasp* of a propositional truth and present naturalistic arguments against his reductive approach to knowledge. We argue that a plausible interpretation of cognitive-science input concerning knowledge—even if one accepts that *knowledge how* is partly propositional—must involve an element of knowing how to act correctly upon the proposition; and this element of knowing how to act correctly cannot itself be propositional.

Keywords: naturalistic epistemology; knowledge how; knowledge that; anti-intellectualism; intellectualism; practical grasp; cognitive science

---

1. Introduction

Our aim in this paper is to use a naturalistic approach to explore Jason Stanley's [1–4] notion of the *practical grasp* of a propositional truth, in light of his intellectualist approach in general. We show that there is more to his notion of "practical grasp" than merely a special kind of relation to a propositional truth, and that this added dimension raises questions about how reducible *knowledge how* is to *knowledge that*. In his investigation of knowledge, Stanley strives to position his theory as being compatible with cognitive science. We will show that his interpretations and conclusions are problematic.

There has arguably been an increased acceptance of naturalism among philosophers, in the sense of a wider acknowledgement of an "all-encompassing" natural world. There is consensus, too, on the importance of taking scientific evidence into account when dealing with philosophical problems, although there is much debate about what exactly such an acceptance ought to involve [5]. An *ontological* naturalism that excludes supernatural entities is nowadays largely uncontroversial. Physicalists [6] claim that physical entities and their arrangements underlie all causal interactions, but it is a much-debated question whether, for example, the mental can be reduced to the physical or not [7]. Influential positions in *methodological* naturalism include, for example, replacement naturalism [8], which claims that epistemology should be replaced with cognitive psychology; substantial naturalism [9], which holds that philosophical problems and issues need to be reformulated in a more strict scientific terminology in order to remain relevant; and cooperative naturalism [10], which claims that philosophy needs to take scientific findings into account where such findings exist. Furthermore, *evolutionary* naturalism [11,12] emphasizes that cognitive agents such as humans are shaped to fit their environment by natural selection. An evolution-based understanding of our cognitive structures and mechanisms can then be seen to replace, succeed, or complement traditional epistemological understandings [13].

We will take a physicalist-inspired approach within cooperative and complementary evolutionary naturalism that views all philosophical questions as being legitimate. The answers need to be compatible with the natural world and the evolved cognitive agents in it, as well as with scientific results. If there is relevant scientific input concerning a particular question or problem at hand, it must be taken into account, and intuitions must give way. We consider this approach both inclusive and potentially fruitful, and so will use it to elucidate the aforementioned influential discussion concerning *knowledge*.

Section 2 briefly discusses Stanley's [1] explanation of how *knowledge how* differs from *knowledge that*. Sections 3–5 evaluate Stanley's basic argument in favour of intellectualism. Section 6 shows how his concept of *knowledge how* counts against elements of his intellectualist argument. Section 7 explores a possible objection to the alternative that we sketch, concluding that—even within Stanley's framework—There must be more to *knowledge how* than the grasp of a proposition. Section 8 presents insights from cognitive science to support our interpretation and discusses how Stanley's position is problematic.

2. The Argument from Knowledge Transfer

A central element in Stanley's [1] theory of knowledge is that practical *knowledge how* can be reduced to propositional *knowledge that*. A general difficulty for any theory that attempts to reduce *knowledge how* to *knowledge that* is that of knowledge transfer. If *knowledge how* is essentially propositional, then it would seem that it ought to be easy to transfer propositionally; yet it is not. Comprehending all the propositional truths about how to swim will not enable one to swim. One cannot learn it from a book. Success takes practice and time. One does not learn it, intellectually; one must learn to do it practically, in a way that seems to point towards some non-propositional proficiency.[1]

To accommodate this practical aspect of *knowledge how* whilst preserving his intellectualist account, Stanley defines *knowledge how* as the practical grasp of a propositional truth: to have mastered a skill is to have grasped a propositional truth in a "practical" way. For an agent to act skillfully, according to Stanley, she must entertain "a practical way of thinking" [1] (pp. 124–130) concerning the true proposition(s) she knows (i.e., her knowledge how). She must grasp the way an action can be performed and be able to perform the action under relevant parameters of normality. She must apply the information she has acquired "in a first-person way" [1] (p. 124): that is, she must relate it to her practical capacities. In this way, *knowledge how* can be seen as propositional, despite possible appearances otherwise.

Given what it is meant to do, the practical grasp of a propositional truth must include the ability to apply that propositional truth practically. If learning to swim means gaining a practical grasp of the propositional truths about swimming, then that practical grasp includes being able to execute the necessary motions described by those propositions: i.e., acting on those propositions in *the right kind of way*. Stanley writes that the inspiration behind his idea can be found in Davidson's philosophy [15]. Davidson added the criterion "in the right way" to avoid the problem of deviant causal chains. Given his causal model, he needed a mechanism that prevents actions coming about in random ways. That said, exactly what constitutes the "right way" was unclear even to Davidson, who wrote [15] (p. 79): "I despair of spelling out … the way in which attitudes must cause actions if they are to rationalize action." Many researchers have drawn attention to this: e.g., Stout [16] (p. 4) mentions "what Davidson famously despaired of spelling out." We think it just as difficult to account for "the right way" in Stanley's account. Stanley's notion of practical grasp has gone through many modifications over the years; we will not treat them all but will focus mostly on his formulation in [1]. Stanley and Williamson [3] write that the agent must entertain a way of Φ-ing under a practical mode of presentation and that [3] (p. 429): "thinking of a way under a practical mode of presentation undoubtedly entails the possession of certain complex dispositions." Stanley [2], [1] builds on this idea

---

[1] For a thorough treatment of this problem, see Glick [14]. He calls it the *sufficiency problem*.

when he develops his view on "ways of thinking" [1] (ch. 4), which involves a "more sophisticated notion of [a] proposition" [1] (p. viii).

There are no uncontroversial views in the literature on the exact meaning of "proposition",[2] and Stanley's is especially hard to pin down. What is clear is that, on his view, propositions are facts and "a proposition is the sort of thing that is capable of being believed or asserted" [1] (p. vii). Thus, there is a duality to his conception: propositions are factual and, therefore, ontological; at the same time, they are something that one can stand in an epistemic relationship to. They appear among a subject's ways of thinking or in different *modes of presentation*.

This conception enables Stanley to argue that skilled action is propositional in character. It is propositional because the skilled agent, in acting, acts on a proposition that she grasps epistemically and can act on because she relates this knowledge to herself in a "first-person kind of way" (see [1] (pp. 85–86, 124–130)). The challenge for the intellectualist is that S can know a proposition p without knowing how to transfer p into action. One could know that w is a way to ride a bike without being able to ride a bike oneself: i.e., there is seemingly a difference between *knowledge that* and *knowledge how*. In his defense of intellectualism, Stanley suggests that *knowledge how* must comprise a special kind of grasp: the practical grasp of a proposition enabling the agent to apply the truth she has acquired to her own agency ("a first person way of thinking" [1] (pp. 85–86)). *Knowledge how* does not consist of a propositional truth *per se* but rather the practical grasp of that propositional truth. The answer to a question about *knowledge how* does not consist of a proposition *per se* but of the "practical grasp" of that proposition, which is what constitutes the skill. When an agent transforms her skill into practice, she does so under a "practical mode of presentation" [2]. The problem with this is that Stanley also argues [1] (pp. 126–128) that knowing how to do something need not mean being capable of doing it: i.e., *knowledge how* need not be the ability to execute a skilled action. Part of our critique is to ask: what *is* the practical presentation in question?

3. Stanley's Intellectualism

The backbone of Stanley's argument for intellectualism is his analysis of embedded questions: i.e., questions that appear in declarative statements or as part of another question. The details of his analysis are complex. To simplify somewhat, he argues that the semantic nature of "how" questions is similar to that of "wh" questions—i.e., who/where/why/when questions—because all conform to the same basic pattern. "Knowing-wh stands and falls together—either they are all species of propositional knowledge, or none of them are" [1] (p. 134). A unified semantic theory requires that *all* answers be propositional. Stanley writes that [1] (p. 141):

> the kinds of philosophical and scientific considerations that would lead us to conclude that knowing how to Φ is not a species of propositional knowledge would also lead us to conclude that knowing where to Φ is not a species of propositional knowledge.

If "how" questions could be answered in a non-propositional way, as anti-intellectualists claim, that would necessarily *also* be true for "wh" questions, which Stanley finds unacceptable: according to him, the answers to these questions often clearly *are* propositional.

Contra Stanley, Devitt [17] argues that one should go ahead and accept that, in many cases, knowing-wh is *not* propositional ([17], as quoted in [1] (p. 134)):

---

[2] It is not our purpose here to give our own view of exactly what a proposition is. Suffice to say, pace Stanley, on our view, S's acting on propositional knowledge requires a kind of knowledge that is not propositional. Propositional knowledge is clearly relevant to action, but we reject the intellectualist claim that *knowledge how* is reducible to *knowledge that*. It is no problem for our view if an agent is unable to verbalize the knowledge she acts on when she acts with skill, but the intellectualist, who takes such knowledge to be propositional (i.e., *knowledge how* is defined as a relation to a proposition), must deliver a convincing account how S can fail to be able to express such knowledge or re-define the term "intellectualist" in an other-than-usual way. We return to this in Section 4 and we explain our own view in more detail in Section 8.

> The foraging desert ant wanders all over the place until it finds food and then always heads
> straight back to its nest. ... On the strength of this competence, we feel no qualms about
> saying that it "knows where its nest is." But to attribute any propositional attitudes to the ant
> simply on the strength of that competence seems like soft-minded anthropomorphism.

Stanley replies [1] (p. 134):

> It should be widely acknowledged that the philosophical and scientific motivations that
> motivate the view that knowing how is not a kind of propositional knowledge also would
> lead one to conclude that many ascriptions of knowing-wh, even ascriptions of knowing
> whether one of several options obtains, do not ascribe propositional knowledge.

"Knowing whether one of several options obtains" is a clear-cut case of propositional knowledge
built on propositional descriptions of the world: either Bogota is the capital of Colombia or it is not.
Stanley takes it as an unwelcome result if a semantic theory of questions says that knowledge-wh
can be non-propositional in character because [1] (p. 134) "languages are remarkably uniform in
their ascriptions of knowing where, knowing when, and knowing whether." He goes on to argue that
knowledge how is likewise expressed in this uniform way cross-linguistically.

At this point, let us set aside further discussion of the details and refer the reader to Stanley.
The bottom line is that he argues that embedded questions demand a unified semantic theory and that
this points in the direction of a *propositional* theory, as many of the questions must be answered in the
form of a proposition.

A possible objection, given Stanley's linguistic focus, is that communication of *knowledge how* does
not work the way he seems to presuppose. In everyday life, "how" questions are not treated in the
same manner as "wh" questions. Imagine what such communication would be like if they were. If one
asks, "how do I swim?" the reply would need to be something like: "place yourself laterally in the
water with your arms extended straight in front of you and your legs behind you. ... First, kick your
legs behind you. Then you just glide for a bit and pull yourself forward with your arms." [18]. If one
asks, "how do I ride a bike?" the reply would need to be something like: "start with one foot on the
ground. Your other foot should be flat on a pedal pointed upwards. Push off, put that foot on the other
pedal, and go! Keep going as long as you can maintain balance." [19].

These questions are common enough; what is rare is the propositionally structured reply: one does
not normally answer these questions linguistically. Such replies seem odd and unhelpful. An exchange
of this sort seldom happens; even when it does, it is of little use in transferring *knowledge how*.

Contrast this with "wh" questions and their replies. If one asks: "where is the nearest gas station?"
the reply might well be something like, "just carry on down this road; there'll be a Texaco on your
right-hand side after half a mile." "When did Susan leave the party?" "She left around ten." "Who is
Barack Obama?" "He served as 44th president of the United States."

Questions and answers of this form are ubiquitous; answers of the "how" form are rare. In the age
of Google and Wikipedia, one can always search the Internet for answers to such "how" questions; but
they do not occur as often in everyday life as "wh" questions do. When—as may rarely happen—they
are answered propositionally, the reply does not *really* answer the question; it cannot because it
does not give the enquirer the *practical competence* she is seeking. Few people ever ask for the best
description of the act of swimming; if one asks, "how do I swim?" that's not what one wants, anyway.
The "how" question is posed with actual performance in mind. Experts typically impart *knowledge how*
by demonstrating it, while the novice imitates what she is shown.[3] Stanley's analysis strikes us as

---

[3] We are simplifying, of course. In practice, there are multiple ways to learn and improve *knowledge how* without recourse to
knowledge that. In addition to imitation, Johansson and Lynøe [20] (pp. 159–161) suggest practicing on one's own, practicing
with a tutor, and learning via "creative proficiency": a practical variation of creative thinking

contrived and not really addressing everyday life. To genuinely know how to do something—to gain that *knowledge how*—one must go ahead and *try* it.

Whereas questions and answers are generally sufficient to gather knowledge in the *wh* domain, they are of little help when it comes to knowledge how to perform a particular task—unless the task is truly simple, like "how do I start the coffee machine?" (to which the reply might be "push the red button"). Such simple questions are quite different from questions concerning how to ride a bike or swim, a point we will return to below. Indeed, we suppose that one could call them camouflaged *knowledge that* questions even though they seem to ask for *knowledge how*.[4] In any case, Stanley's interest is not in these simple cases but in complex activities like swimming or riding a bike.

Though Stanley puts a great deal of weight on answering questions about such activities, he does not claim that people ask and answer "how" questions in the way we have discussed. Instead, he has a special sense of answering questions in mind.

4. Answering Questions

Building on his earlier work with Williamson, Stanley holds that *knowledge how* always amounts to knowing the answer to a question; but this does not mean that all such knowledge consists in the ability to verbalize one's answers, because [2] (p. 214) "knowing how to ride a bicycle involves knowledge of a distinct proposition than does knowing how to explain how to ride a bicycle."[5] One's ability to answer a question can be latent, in which case one simply acts, based on implicit facts; as Stanley puts it [2] (p. 214): "only grasping a way to ride a bicycle is required to know how to ride a bicycle." The cycle-riding agent has, in a practical way, grasped the requisite propositions for riding a bike: i.e., she knows of some means, *w*, such that *w* is a way to ride a bicycle. Her *knowledge how* consists in an ability to answer certain questions in principle, not necessarily in practice.

It is important for Stanley that, even though knowledge how to Φ is the same as knowing a fact about Φ, the intellectualist does not demand from the agent that she can explain how to Φ. The only requirement is that she grasps a way in which Φ can be performed, though Stanley understands that one might think otherwise [2] (p. 214): "perhaps the very fact that the intellectualist defines knowing how in terms of propositional knowledge suggests that someone who knows how to do something must be able at least to *express* her propositional knowledge." The point is that the agent need only be able to express that knowledge *somehow*, perhaps with demonstratives, not necessarily explain it verbally: a weaker claim than the one might otherwise think Stanley is making.

Stanley explicates his point by offering the young Mozart who, having been asked, "how do you do it?" points to one of his masterpieces while writing it, saying [2] (p. 214): "this is how I can do it." Of course, one could object that this does not answer the question of how to compose a masterpiece. An answer should be informative, and Mozart's "answer" is not illuminating at all—not so much because he cannot verbalize himself as because he is seemingly unable to give any satisfactory answer. It seems counterintuitive to say that he possesses an ability to answer the "how" question presented to him. Nevertheless, Stanley claims that Mozart's pointing at his masterpiece, while composing it, is an expression of *knowledge how*. He justifies this by asking one to consider what words should count as expressing *knowledge how*. If *all* words should count—and Stanley believes they should—then, Stanley says, demonstrative expressions like "this" count, and Mozart has successfully expressed his *knowledge how*.

---

[4] Arguably, many everyday *knowledge how* questions, and basically all those one expects to be answered verbally, are concealed *knowledge that* questions: e.g., "how do I get to the bus stop?" and "how do you play chess?" are both asked in a "how" manner but are really requests for *knowledge that*.

[5] Stanley quotes Fodor: "There is a real and important distinction between knowing how to do a thing and knowing how to explain to do that thing. But that distinction is one that the intellectualist is perfectly able to honor … The ability to give explanation is itself a skill—a special kind of knowing how which presupposes general verbal facility at the very least. But what has this to do with the relation between knowing how and knowing *that*? And what is there here to distress an intellectualist?" (quoted in Stanley [2] (p. 213).

Stanley does not seem much concerned with how the recipient interprets Mozart's gesture. What matters is that Mozart has a way to express his knowledge that Stanley's intellectualism can capture. Stanley goes further: the intellectualist need not hold that an agent is able to express the proposition that represents her *knowledge how* in words at all, not even using demonstratives. Her *knowledge how* need be nothing more than a state implicated directly in action. Stanley writes [2] (p. 215):

> The southpaw is winning on points. But then the expert boxer adjusts and starts boxing in a particular way that is the best way to fight against a southpaw. The announcer, pointing at the way in which the expert boxer is fighting, utters: "He knows that that's the best way to beat a southpaw." *The announcer's knowledge-ascription is quite explicitly a true ascription of knowledge-that.* Furthermore, it is true *whether* or not the boxer is able to verbalize his knowledge of the way in question of boxing against a southpaw in non-indexical terms, non-demonstrative terms.

The point Stanley wants to make here is that it is not a requirement on his intellectualist account that *S* can verbalize his *knowledge how*. The expert boxer knows how to beat a southpaw without being able to verbalize this knowledge. This, he says, is analogous to the fact that one can know multiple different shades of a colour without being able to express this knowledge [2] (p. 215).

We have two responses to this. First, that an observer—the announcer in this case—can have propositional knowledge about another person's *knowledge how* does not imply that the latter knowledge is not of a different, more *practical* sort. Second, is it genuinely possible to know the answer to a question, in the sense required by Stanley's account, if one is unable to verbalize one's *knowledge how*? As we argue throughout, an intellectualist account which accepts a "mute" grasp of propositions is highly problematic. It runs the risk of ending up in precisely the type of conception of *knowledge how* that it seeks to avoid. What seems to constitute a practical mode of presentation for Stanley involves some kind of expertise irreducible to the proposition in itself. Moreover, Stanley's conception of *knowledge how* seems very distant from how research suggests that cognitive processes work.

Stanley's expert boxer has grasped certain boxing truths, expressed through his change of tactics. We have a problem with this. If, as Stanley claims, knowledge always involves grasping a proposition, and this is meant to capture how actions are informed by intelligence, it seems to us a proper demand that the agent should be able to give a more informative expression to the proposition she is meant to grasp. If she cannot do so, it is hard to see in what sense she grasps a proposition. Stanley's "mute" grasp of a proposition strikes us as mysterious, yet Stanley offers nothing to dispel the mystery. Instead, he focuses on shared characteristics of "how" and "wh" questions.

5. Wh-Questions

Remember that, for Stanley, two things are key: *knowledge how* consists in knowing the answer to a question, while *knowledge how* should fit within a more general account of knowing answers to questions. Toward that end, he seeks to set out for his readers the similarities between knowledge how and knowledge-*wh* and so between "how" and "wh" questions. He believes that knowledge ascription is always done in a similar way, comprising the ability to answer a "how" or a "wh" question. Just as one says things like "Hannah knows *where* her bike is", "Hannah knows *why* her bike has a flat tire", and "Hannah knows *when* she parked her bike in the garage", so one says things like "Hannah knows *how* to ride a bike." Of course, one *could* give each of these ascriptions its own account: one for *knowledge where*, one for *knowledge why*, one for *knowledge when*, and so on. What is striking, though, is that all these ascriptions have a similar semantic structure [1] (p. 36): "it is a stable cross-linguistic fact that most of the sentences ... are translated with the same verb used in translations of sentences of the form 'X knows that p'", and "the same word 'know' occurs in all of these constructions." For Stanley, that means they represent a single, unified phenomenon. He writes that [1] (p. 37): "the fact that we do not employ different words for these notions suggests they are at the very least intimately related concepts."

Rumfitt [21] criticizes this line of reasoning. He notes that French everyday language contains several distinct terms for knowledge how: e.g., *savoir faire*. Such terms do not seem to fit the pattern Stanley prescribes.

Stanley replies that the apparent problem is nothing more than a matter of Gricean conversational implicature: while many languages spell out embedded questions in full in accordance with his model, others do not. Their omission of the question word is guided by Grice's maxims [1] (p. 141, *emphasis added*):

> It is clear that in a language in which it is possible to drop the overt question word in expressions of knowing how, Grice's maxim of manner predicts that one ought to drop the question word. But there is no language known to me where the propositional verb together with the bare infinitive means *knowing where. So the fact that in many languages ascriptions of knowing how do not superficially appear to take the form of an embedded question should not lead us to analyze them as relations to activities. So doing would lead to an unwarranted asymmetry between states of knowing how to do something and states of knowing when to do something and where to find something, asymmetries that all parties to the debate about the nature of practical knowledge should reject.*

Not only does Stanley seem to think he can explain linguistic developments in the French language by appeal to Grice, he also claims that interpreting *knowledge how* as "relations to activities" would lead to an "unwarranted asymmetry" between states of knowing how and other states of knowing-wh. When Stanley says that *knowledge how* is propositional, he really does mean that it is propositional on a par with knowing where the nearest gas station is; but this raises a difficulty.

6. Ability to Execute a Skilled Action

For a unified theory of embedded questions to be possible in the way Stanley wants, the answer to a "how" question, as an expression of *knowledge how*, must consist of a proposition. As we argued in Section 2, *knowledge how* is the ability to execute a skilled action—noting that, for Stanley, it is not a requirement of *knowledge how* that the agent actually is able to execute the skill; an abstract ability suffices. (We will say more on that below.)

The problem is that, by virtue of the argument from knowledge transfer, it cannot be the mere grasp of a proposition that constitutes *knowledge how*, because *knowledge how* cannot be transferred propositionally. *Knowledge how* must comprise a certain practical kind of grasp. *Knowledge how* does not consist of a propositional truth *per se* but rather the practical grasp of that propositional truth. It follows that the answer to a question about *knowledge how* does not consist of a proposition *per se* but of the practical grasp of that proposition, which is what constitutes the skill. If that is so, it is not the case, *pace* Stanley, that the answers to embedded questions have uniform structure?[6]

If we are right, this weakens Stanley's case for intellectualism. Despite what he claims, his theory seems unable to provide a unified semantic theory, at least when it comes to embedded questions. More tellingly, we believe we have identified an implicit component of genuinely non-propositional knowledge lurking within his theory: what he calls the practical grasp of the proposition. That grasp constitutes the ability to act upon the proposition by performing a skilled action; but does not this ability then constitute a competence in itself, independent from the proposition? If so, is it not this competence that really constitutes *knowledge how*?

---

[6] We are aware it may seem odd to say that *knowledge how* should consist in a propositional truth, but this is a consequence of Stanley's theory. Remember he holds that *knowledge how* must consist of an answer to a question. Given that he wants a unified semantic theory of questions, the answer to a question—including the "how" question—must consist of propositions.

7. The Ability Objection

There is a potential problem with our position. At one point, Stanley argues [1] (pp. 126–128) that knowing how to do something need not mean actually being capable of doing it: contrary to what we have claimed, *knowledge how* need not consist of the ability to execute a skilled action. If it does not, then the difficulty we have outlined in Stanley's account threatens to dissolve.

Stanley's claim rests on three examples, two of which are taken from Ginet [22]. Ginet's eight-year-old son is not strong enough to lift a certain box; nonetheless, Ginet, and Stanley, say that he must be said to know how to lift it, because he knows how to lift boxes in general. Stanley writes [1] (p. 128): "Ginet's son knows how one *could* lift one hundred pounds off the floor … " The second example, also from Ginet, concerns an expert skier who is unable to ski down a hill because of stomach cramps. Certainly, he knows *how* to ski down the hill, even though he cannot execute the ability at the moment. The third example, taken from Stanley and Williamson [3] (p. 416), concerns a concert pianist who loses both arms. Obviously, she can no longer play the piano, but, given her many years of practice, she still knows how to do so. Stanley believes that, together, these examples support the view that *knowledge how* need not entail the ability to execute a skill but rather takes the form of more abstract knowledge—thereby bolstering the case for his intellectualism.

We have two considerations in reply. First, how do the examples relate to what we have otherwise established? In the argument from knowledge transfer (Section 2), we conclude that *knowledge how* cannot consist in grasping a proposition in exactly the same sense as grasping a proposition theoretically since merely knowing how to do something theoretically does not enable one to do it. If *knowledge how* involves grasping a proposition at all, it must be a special kind of grasping.

It seems odd, then, that this special kind of grasping need not enable one to execute the relevant skill, and not just because of immediate circumstances such as being too young, having stomach cramps or losing one's arms: in other words, it seems odd that one can grasp a proposition "practically" without being able to act on it. The very reason why Ryle [23] separated *knowledge how* from *knowledge that* in the first place was to make room for the reality that is the practical execution of actions. What Stanley's intellectualism risks leaving one with is a notion of *knowledge how* that is practical only in name.

This leads to our second consideration. All three of Stanley's examples have the same basic structure: an agent faces a task that she cannot perform, but which she could perform under *other* circumstances. The eight-year-old can lift boxes, just not this box—until he is older. The skier can still ski, just not until the stomach cramps pass. The pianist knows how to play and *could* play if she, for example, got robotic prostheses. All the examples allow a hypothetical scenario wherein the agent is able to execute the skill.

Stanley writes that these agents possess the relevant *knowledge how* but are unable to execute it at a particular time. The eight-year-old certainly *knows* how to lift the heavy box; he just cannot do it because he is not strong enough. Stanley is stressing his brand of a more abstract form of *knowledge how*. He concludes further that *knowledge how* need not be tied to execution but takes a more abstract form. However, could not the reason one has to ascribe *knowledge how* to these agents instead be the existence of the hypothetical scenarios? If Stanley had not stressed that Ginet's son knows how to lift boxes and simply described an eight-year-old who cannot lift a certain heavy box, would one still say that the child knows how to lift it? At the least, intuitions may differ.

If one leaves oneself open for the kind of *knowledge how* ascriptions that Stanley wants, then one must also allow that someone can know how to play the piano, even though they have never tried. That person could know how, because she knows that, to play the piano, one "just" needs to hit the keys in the correct combinations and sequence using the correct pressure and timing—just as the eight-year-old knows that, to lift a heavy box, one "just" needs to grab hold of the edges and stand straight. We take this to be a biologically implausible consequence. If one is to be said to know how to play the piano, it seems reasonable to demand that one knows how to use it to produce actual, and not hypothetical, music.

8. The Cognitive Science of Knowledge

While Stanley's main focus is to build a theory of knowledge based on linguistic arguments and reflections, he invokes cognitive-science results to strengthen his case [1,4]. Let us take a closer look at some of the relevant findings of cognitive science in light of the considerations we have raised against Stanley. We believe that the naturalistic input which Stanley invites speaks strongly against his position.

The cognitive-psychological dual process theory offers a canonical rendering of cognitive processes that divides the mind into a bottom-up reflexive form (System 1, Type 1 processes) and a top-down reflective form (System 2, Type 2 processes). We wish to use this influential theory as a heuristic framework to inform our account of knowledge.

According to dual process theory, the reflexive form is non-conscious and governs automatic processes, whereas the reflective form is conscious and governs reflective processes [24–28]. Though interpretations of the available evidence vary, there is nonetheless strong evidence that this framework picks out interesting features of human cognition [27,29]. The framework is further supported by how well it coheres with leading memory-systems theories on a lower cognitive-psychological level of analysis. Both Tulving's seminal work on long-term memory [30,31] and Baddeley's on working memory [32] fit generally well with the dual process theory. Procedural memory governs perception, motor functions, and procedural knowledge. Semantic memory governs pattern recognition, categorization, and conceptual knowledge. Episodic memory governs self-awareness and our remembrance of events. Finally, working memory governs several functions correlated with reflection, attention, and executive control. Procedural (non-declarative) long-term memory roughly coheres with System 1, while episodic (declarative) memory together with working memory cohere with System 2.[7]

Both forms of processes are crucial for cognition and fill important, yet distinct, roles. Bottom-up reflexive processes are context-specific whereas top-down reflective processes offer generalizability by sometimes inhibiting reflexive processes. Even though Type 1 processes can appear reflexive and instinctive, they are shaped by repeated experience. Reflective Type 2 processes are more flexible, having the capacity to quickly adjust quickly to new circumstances [33] (pp. 132–133).[8]

Briefly, thoughts and intentions (*knowledge that*) involve working memory and episodic memory based in frontal cortical regions, consisting of homotypical neural cells. Motoric competence (*knowledge how*), on the other hand, involves procedural memory based in the primary motor cortex and the cerebellum, consisting of large agranular neural cells [27,30–32,36–38]. It is plausible to interpret these findings as supporting a distinction between *knowledge how* and *knowledge that*: the two knowledge forms involve different functions, brain regions, and cell types. Stanley [2,4] argues against such an interpretation but does not, in our view, provide any convincing evidence or arguments for doing so.

It is indeed possible to improve one's skills—one's *knowledge how*—by gaining relevant propositional *knowledge that*; but it is only after repeated actual performances of actions that relevant neural connections are established and *knowledge how* developed. Stanley [2] (pp. 155–159) discusses these matters but does not properly acknowledge how primarily non-conscious and reflexive motoric processes govern an agent's ability to act—not propositional aspects [39].

No matter how people may *ascribe* knowledge, the facts of the matter remain: *knowledge how* involves very different processes than *knowledge that*. Empirical evidence suggests strongly that procedural memory and *knowledge how* are non-propositional [40,41].

Returning to the earlier examples, lifting boxes is such an elementary action that one will entrain the necessary neural connections almost by default. Both downhill skiing and piano playing, however, involve motoric competencies that demand specific practice in order to be developed. If such

---

[7]   It is less clear how semantic long-term memory ought to be positioned in this multi-level picture, and so we will presently leave it as an open question.

[8]   For critical views see, e.g., [34,35].

practice is absent—if only propositional *knowledge that* is present—*knowledge how* will not come about. The important thing is not what propositions an agent knows, but what actions she has repeatedly performed, which have the potential to lead her to develop *knowledge how*.

9. Conclusions

We have attempted to argue that *knowledge how* must comprise more than the mere grasp of a proposition even if one otherwise accepts Stanley's intellectualism. We have done this by describing an implicit conflict between what we have called the argument from knowledge transfer and Stanley's argument from embedded questions: one that is revealing of certain characteristics of Stanley's "practical grasp" of a proposition that he downplays. At heart, our objection to Stanley is simple: even if *knowledge how* is propositional, it must involve an element of knowing how to act correctly upon the proposition; and this element of knowing how to act correctly cannot itself be propositional. Thus, *knowledge how* involves an irreducible non-propositional element and cannot be reduced to *knowledge that*. As Ryle [23] (p. 28) put it: "to be intelligent is not merely to satisfy criteria, but to apply them." In accordance with our naturalistic stance, we have looked for input from cognitive science and found it offering convincing support for our non-reductive approach.

Author Contributions: Both authors have contributed to: Conceptualization, Methodology and Writing—Original draft preparation, Review and Editing. All authors have read and agreed to the published version of the manuscript.

Funding: This research received no external funding.

Acknowledgments: Thanks to Santiago Amaya, Dan Egonsson, Olav Gjelsvik, Ingvar Johansson, Katarzyna Paprzycka, Joel Parthemore, Björn Petersson, Wlodek Rabinowicz, Andreas Engh Seland and our anonymous reviewers for valuable feedback on this text. Sections 2–7 based on C.V. Felix's doctoral thesis *Slips, Thoughts and Actions* (2015).

Conflicts of Interest: The authors declare no conflict of interest.

References

1. Stanley, J. *Know How*; Oxford University Press: Oxford, UK, 2013.
2. Stanley, J. Knowing (how). *Nôus* **2011**, *45*, 207–238. [CrossRef]
3. Stanley, J.; Williamson, T. Knowing-how. *J. Philos.* **2001**, *98*, 411–444. [CrossRef]
4. Stanley, J.; Krakauer, J.W. Motor skill depends on knowledge of facts. *Front. Hum. Neurosci.* **2013**, *7*, 1–11. [CrossRef] [PubMed]
5. Rysiew, P. Naturalism in Epistemology. *The Stanford Encyclopedia of Philosophy*. Spring 2017 Edition. Available online: https://plato.stanford.edu/archives/spr2017/entries/epistemology-naturalized/ (accessed on 2 March 2020).
6. Smart, J.J.C. Sensations and brain processes. *Philos. Rev.* **1959**, *68*, 141–156. [CrossRef]
7. Stoljar, D. Physicalism. *The Stanford Encyclopedia of Philosophy*. Winter 2017 Edition. Available online: https://plato.stanford.edu/archives/win2017/entries/physicalism/ (accessed on 2 March 2020).
8. Quine, W.V.O. Epistemology Naturalized. In *Ontological Relativity and Other Essays*; Columbia University Press: New York, NY, USA, 1969; pp. 69–90.
9. Churchland, P.S. Mind-brain reduction: New light from the philosophy of science. *Neuroscience* **1982**, *7*, 1041–1047. [CrossRef]
10. Kornblith, H. *Knowledge and Its Place in Nature*; Oxford University Press: Oxford, UK, 2002.
11. Sellars, R.W. *Evolutionary Naturalism*; Open Court Publishing Company: London, UK, 1922.
12. Campbell, D.T. Evolutionary epistemology. In *The Philosophy of Karl R. Popper*; Schilpp, P.A., Ed.; Open Court: LaSalle, IL, USA, 1974; pp. 412–463.
13. Bradie, M.; Harms, W. Evolutionary Epistemology. *The Stanford Encyclopedia of Philosophy*. Spring 2017 Edition. Available online: https://plato.stanford.edu/archives/spr2017/entries/epistemology-evolutionary/ (accessed on 2 March 2020).
14. Glick, E. Practical modes of presentation. *Nôus* **2015**, *49*, 538–559. [CrossRef]
15. Davidson, D. Psychology as philosophy. In *Essays on Actions and Events*, 2nd ed.; Clarendon Press: Oxford, UK, 2001.

16. Stout, R. *Things That Happen Because They Should*; Oxford University Press: Oxford, UK, 1996.
17. Devitt, M. Methodology and the nature of knowledge-how. *J. Philos.* **2011**, *108*, 205–218. [CrossRef]
18. Fang, A. How to Swim the Breaststroke. wikiHow. 2019. Available online: http://www.wikihow.com/Swim-the-Breaststroke (accessed on 2 March 2020).
19. wikiHow Staff. How to Ride a Bicycle. wikiHow. 2020. Available online: http://www.wikihow.com/Ride-a-Bicycle (accessed on 2 March 2020).
20. Johansson, I.; Lynøe, N. *Medicine and Philosophy: A Twenty-First Century Introduction*; Ontos Verlag: Frankfurt, Germany, 2008.
21. Rumfitt, I. Savoir faire. *J. Philos.* **2003**, *100*, 158–166. [CrossRef]
22. Ginet, C. *Knowledge, Perception, and Memory*; Dordrecht Reidel: Boston, MA, USA, 1975.
23. Ryle, G. *The Concept of Mind*; Chicago Press: Chicago, IL, USA, 2002.
24. Tversky, A.; Kahneman, D. Judgement under uncertainty: Heuristics and biases. *Science* **1974**, *185*, 1124–1131. [CrossRef]
25. Tversky, A.; Kahneman, D. Extensional vs. intuitive reasoning: The conjunction fallacy in probability judgment. *Psychol. Rev.* **1983**, *90*, 293–315. [CrossRef]
26. Kahneman, D. *Thinking Fast and Slow*; Farrar, Straus and Giroux: New York, NY, USA, 2011.
27. Evans, J.S.B.T.; Stanovich, K.E. Dual-process theories of higher cognition: Advancing the debate. *Perspect. Psychol. Sci.* **2013**, *8*, 223–241. [CrossRef] [PubMed]
28. Lizardo, O.; Mowry, R.; Sepulvado, B.; Stoltz, D.S.; Taylor, M.A.; Van Ness, J.; Wood, M. What are dual process models: Implications for cultural analysis in sociology. *Sociol. Theory* **2016**, *34*, 287–310. [CrossRef]
29. Frankish, K. Dual-process and dual-system theories of reasoning. *Philos. Compass* **2010**, *5*, 914–926. [CrossRef]
30. Tulving, E. Memory and consciousness. *Can. Psychol.* **1985**, *26*, 1–12. [CrossRef]
31. Tulving, E. Episodic memory: From mind to brain. *Annu. Rev. Psychol.* **2002**, *53*, 1–25. [CrossRef]
32. Baddeley, A.D. *Working Memory, Thought and Action*; Oxford University Press: Oxford, UK, 2007.
33. Plotkin, H.C. *Darwin Machines and the Nature of Knowledge*; Harvard University Press: Cambridge, MA, USA, 1993.
34. Gigerenzer, G.; Regier, T. How do we tell an association from a rule: Comment on Sloman. *Psychol. Bull.* **1996**, *119*, 23–26. [CrossRef]
35. Keren, G.; Schul, Y. Two is not always better than one: A critical evaluation of two-system theories. *Perspect. Psychol. Sci.* **2009**, *4*, 533–550. [CrossRef]
36. Kandel, E.R.; Schwartz, J.H.; Jessell, T.M.; Siegelbaum, S.A.; Hudspeth, A.J. *Principles of Neural Science*, 5th ed.; McGraw-Hill, Health Professions Division: New York, NY, USA, 2013.
37. Allen, T.A.; Fortin, N.J. The evolution of episodic memory. *Proc. Natl. Acad. Sci. USA* **2013**, *110*, 10379–10386. [CrossRef]
38. Buzsáki, G. *Rhythms of the Brain*; Oxford University Press: Oxford, UK, 2006.
39. Balleine, B.W.; O'Doherty, J.P. Human and rodent homologies in action control: Corticostriatal determinants of goal-directed and habitual action. *Neuropsychopharmacology* **2010**, *35*, 48–69. [CrossRef]
40. Aarts, H.; Dijksterhuis, A. Habits as knowledge structures: Automaticity in goal-directed behavior. *J. Personal. Soc. Psychol.* **2000**, *78*, 53–63. [CrossRef]
41. Noë, A. Against intellectualism. *Analysis* **2005**, *65*, 278–290. [CrossRef]

 *philosophies*

Article

De Libero Arbitrio—A Thought-Experiment about the Freedom of Human Will

Johannes Schmidl

Save Energy Austria, Franz Josefs Kai 13/12-13, 1010 Vienna, Austria; johannes.schmidl@chello.at

Received: 8 January 2020; Accepted: 11 February 2020; Published: 16 February 2020

Abstract: The discussion of whether or not humans are able to act freely is ongoing, even though, and precisely because, technical methods for detecting the physical state of the brain are constantly improving. The brain as a physical–chemical object seems to be pre-determined by its physical and chemical states, while at the same time human consciousness gives the impression of being able to decide subjectively and freely on its own. Determinists claim that this free decision is just a form of misinterpretation of an epiphenomenon and that the alleged "free decision" has actually been determined by the physical state of the brain before the human subject gives the impression of being able to decide freely. The basis for this is a set of experiments, the first of which was specified by Benjamin Libet. Determinism, as the philosophical position that all events are entirely determined by previously existing causes, in principle enables the existence of a perfect predictor. In this paper, a thought-experiment is introduced which demonstrates that a subjective consciousness can break any forecast about its physical state, independently of the method of its detection, and, consequentially, to refute claims about its purely deterministic role. The thought-experiment picks up on an idea of the philosopher Alvin I. Goldman. Logically, the proof follows the path of a 'reductio ad absurdum'.

Keywords: thought-experiment; libero arbitrio; freedom of will

---

1. Introduction

In an interview for "New Scientist" in February 2011, biological novelist Christian de Duve stated that the cost of human success on earth is the exhaustion of natural resources, leading to energy crises, climate change, pollution and the destruction of our habitat. If we continue in the same direction, humankind is heading for some frightful ordeals, if not extinction. He believes an inherent selfishness in human nature is responsible for this behaviour passed on through human genes. [1]

This somewhat fatalistic statement raises an old philosophical question: are human beings able to act according to free will or are they pre-determined—pre-programmed—by the physical state of their brain, which is based on their genetic traits.

First of all, the set-up, results, and two main conclusions of the famous Libet-experiment [2] have to be described in order to be able to shed light on the context of the thought-experiment:

The set-up of the experiment involves EEGs being attached to subjects' brains, monitoring when their brains indicate a decision is made and comparing this to when subjects indicate being aware of having made a decision. The results show that the pre-conscious brain indicates a reliably predictive "readiness potential" signal approximately 0.33 s prior to test subjects reporting an awareness of having made a decision on their own. The two main interpretations are:

- First, as the readiness potential reliably predicts the subsequent decision, Libet concluded that the brain has decided prior to having become aware of consciously making a decision. This implies that the conscious awareness of having made a decision is an epiphenomenon, which undermines the free will notion that volition is informed by conscious election.

- Second, Libet was nonetheless of the opinion that this study does not rule out the possibility that individuals retain a "veto" option, that is to say they can override what the brain delivers, if they so choose, and veto the brain's decision just prior to the moment of enacting it.

More recent work has established, first, that when test subjects were asked to press either a left-hand or right-hand lever, analyses of their brain states reliably predicted which hand would be chosen up to six seconds prior. Secondly, the readiness potential itself has been debunked as a sort of mathematical noise of the equations used to measure brain activity.

This paper links its interpretation of the thought experiment with Libet's assertion that the conscious agent retains a veto power. We shall assume test subjects would wish to falsify such a prediction every now and then, supporting the existence of free human will.

2. A Thought-Experiment in Favour of Free Human Will

In preliminary matters, we need one axiom:

A system is considered to be determinate, as long as its state and therefore resulting future can in principle be known in advance. For the following thoughts, it is sufficient if the state and the further dynamic development of the system can be known at least two seconds or more into the future.

We start with the system in Figure 1. The system consists of the following elements:

- a human being, indicated in the figure by its brain and its eye;
- a detecting probe with related software and representation on a screen, which allows for the full detection and representation of the complete physical/chemical/biological state of the brain;
- an observer, usually a human being. The observer can also be placed outside the frame of Figure 1, meaning that it is not decisive that they are part of the system Figure 1, which is considered to be deterministic, or not;
- a simple decision-process for the human being, indicated by the two cups A and B;
- and the environment of the experiment, which can also be part of the deterministic system.

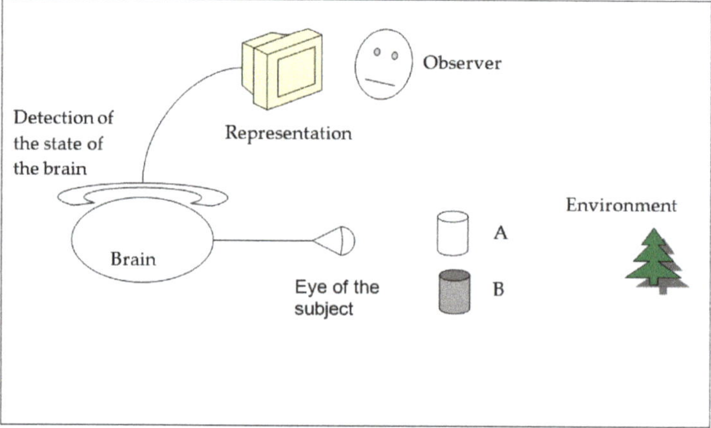

Figure 1. A system consisting of a human who has to decide whether to drink a cup of pure water (A) or one of wine (B). Only the brain and eye of the human being is sketched. The brain of the human is fully detected; its state of being is represented on a screen. If the way the human brain works is predetermined, the observer will know in advance if the human will choose possibility A or B.

The whole system in Figure 1 is considered to be pre-determined, a classical physical system without in-deterministic influences. The brain of the human being produces electrochemical results which "somehow" are related to subjective perception. Instantaneously, the whole objective system of the "brain" with all its signals is detected by a probe, interpreted by a computer and is displayed to an observer. The "somehow"-production of subjective impressions as well as the details of the detecting probe and the representation of the state of the brain, are not the focus of this paper.

The observer can, as a consequence of detection, interpretation and representation, know in advance what the electrochemical system "brain" intends to do, when its subjective correlate, the human consciousness, is confronted with a simple decision. The human has to decide between two options with almost the same level of priority, meaning that both are more or less equal for them. In our example, the human has to decide whether to drink from cup A which contains pure water, or from cup B which contains red wine. This is also the principal idea of Libet's experiment.

The system "brain" is a determined physical/chemical/biological system, as are the rest of all elements in Figure 1. In Figure 1, the brain would make the decision prior to the human's subjective perception where it actively decides between possibilities A and B. Therefore, the observer would know: "you decide to choose option A" and the human would follow. Everything remains in the predetermined world.

Now, let us change the situation in Figure 1 slightly to that in Figure 2.

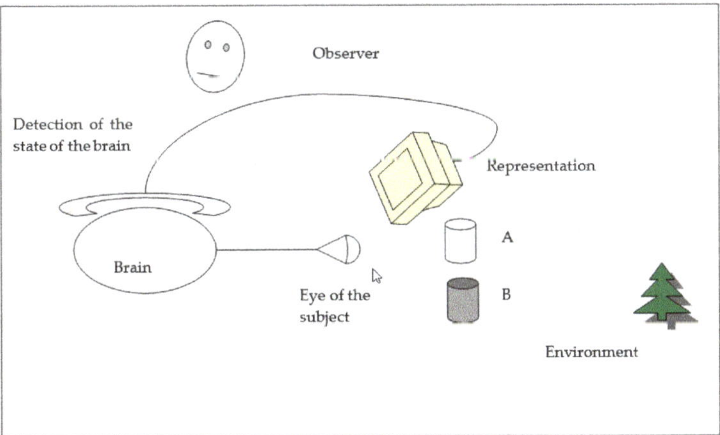

Figure 2. This is the same system as in Figure 1, except for one difference: now the human actor who has to decide whether to take cup A or B can simultaneously watch a screen which would represent to himself/herself the inner state of his/her brain. The human actor would know at least about two seconds in advance which of the two possibilities, A or B, his/her brain would select. Due to his/her consciousness, however, the human actor is now able to refute any predetermined decision of his/her brain.

We turn the screen representing the inner "physical" state of the human brain around so that the eye of the human being, the brain—which is being detected and observed—is able to watch it (Figure 2). Again, the whole system, as a hypothesis, is considered to be determined.

In this case the human being—again sketched as a brain and eye in Figure 2—would learn about two seconds in advance which decision it is about to choose (A in our example). This again happens prior to the human's subjective perception that it has actively decided between possibilities A and B. Is the human being now obliged to choose alternative A, as the representation of the inner physical state of his/her brain tells them via the screen that it is about to choose alternative A? Definitely not.

In Figure 2 there is unquestionably something placed in the system which is able to refute any forecast of the deterministic system. Whenever the chain along brain-detection-representation and observation (b-d-r-o) mirrors a certain result, something in between is able to prove that this very result can also be different.

This "something in between" is the free, resistive will (libero arbitrio). The hypothesis represented in Figure 2, that the whole system is a pre-determined system, is disproved.

It is important to keep the following in mind:

1. Even if there were to be a logical element between the representation of the inner state of the brain and the brain itself (meaning somewhere "within" the human being), which were able to produce the output non-A from input A (or carry out any other logical operation), this logical element itself would again be part of the pre-determined system within the frame of Figure 2. Therefore, its state and its future development would both be known in advance. The same would occur if the brain itself were to merely produce this logical element. In a classical system, there is no principal hurdle against forecasting the whole state and the whole future development of the system in Figures 1 and 2, deciding whether a human would select alternative A or B.

2. Both systems within the frames of Figures 1 and 2 are classical systems in the way that quantum mechanical effects and related uncertainties do not play a role in them.

3. Conclusions

The critical element in Figure 2, which holds the key, is the brain with its resistive will. This brain is both an object and at the same time a related subject. Simultaneously, the subject observes itself as an object. Determinists state that the object "somehow" produces the subject as a correlated epiphenomenal side-effect. The constellation of Figure 2 shows, however, that the subject can disprove the state of its relating object by refuting its predicted future. It is able to disprove decisions that the object seems to have taken.

If anything that is predicted about the system can be refuted, the system turns into a non-predictable one.

In 1970, the philosopher Alvin I. Goldman [3] proposed a thought-experiment of a perfectly informed, accurate deterministic predictive system, predicting that a subject will perform a certain action (say, raise her hand) at time *t*. If the subject were aware of that prediction, they can simply refuse to raise their hand at *t*. The thought experiment presented here is designed to support the same idea, by having the deterministic predictive system directly monitor brain and environment of a subject and feed the prediction back to them. It places the bases for the deterministic prediction inside the test subject's own brain and provides the test subject with what counts as neurological biofeedback, together with deterministic knowledge about their own brain, thereby bringing the thought experiment much closer to home for our intuitions about ourselves and our brains in a way that relates more directly to Libet's veto power claim.

4. Further Questions and Remarks

The hypothesis assumes, for the sake of its reductio ad absurdum approach, that the system is reliably deterministically predictive. If so, then the system would be able to know whether the test subject wishes to rebut its deterministic predictions and will integrate that information into its predictive processing algorithms. Again, it would mirror these predictions of the inner state of their brain to the human actor, and the human actor would again be able to react with obeying or refuting this prediction. Eventually the potential reaction-time between prediction, presentation and the reaction would become too short for a resistive reaction on the part of the human actor. Therefore, we must define a threshold period for the game of prediction and reaction of, say, half a second or two seconds.

Furthermore, the hypothesis assumes that a free human actor is characterized by the desire to now and then falsify reliable deterministic predictions about their forthcoming behaviour. There is no empirical evidence for that, and it would be hard to find one that avoids the trap of tautology. Rather, this resistive desire is taken as a definition for a free will.

It should be noted that this thought-experiment is akin to the famous Newcombe's paradox with a perfect predictor. If we were to redesign Newcombe's paradox in the way that a perfect predictor would, in addition to our knowledge of the content of the box containing a million dollars or nothing, predict the complete physical/chemical/biological state of the brain of the human player and mirror it back to this very player (as indicated in Figure 2), it would be possible for this player to win the million dollars.

Funding: This research received no external funding.

Acknowledgments: The author wants to thank colleagues for their valuable time for discussion of the thought-experiment, especially Dieter Bandhauer, Lutz Musner, Andreas Obrecht, and Walter Peissl.

Conflicts of Interest: The author declares no conflict of interest.

References

1.   Witchalls, C. Biology Nobelist: Natural Selection Will Destroy Us. Available online: https://www.newscientist.com/article/mg20928015-400-biology-nobelist-natural-selection-will-destroy-us/ (accessed on 2 January 2019).
2.   Libet, B. Unconscious cerebral initiative and the role of conscious will in voluntary action. *Behav. Brain Sci.* **1985**, *8*, 529–566. [CrossRef]
3.   Goldman, A. *A Theory of Human Action*; Prentice-Hall: Englewood Cliffs, NJ, USA, 1970.

 *philosophies*

Article

Spurious, Emergent Laws in Number Worlds

Cristian S. Calude [1,*] **and Karl Svozil** [2]

1   School of Computer Science, University of Auckland, Private Bag 92019, Auckland 1010, New Zealand
2   Institute for Theoretical Physics, Vienna University of Technology, Wiedner Hauptstrasse 8-10/136, 1040 Vienna, Austria; svozil@tuwien.ac.at
*   Correspondence: cristian@cs.auckland.ac.nz

Received: 12 February 2019; Accepted: 8 April 2019; Published: 16 April 2019

Abstract: We study some aspects of the emergence of *lógos* from *xáos* on a basal model of the universe using methods and techniques from algorithmic information and Ramsey theories. Thereby an intrinsic and unusual mixture of meaningful and spurious, emerging laws surfaces. The spurious, emergent laws abound, they can be found almost everywhere. In accord with the ancient Greek theogony one could say that *lógos*, the Gods and the laws of the universe, originate from "the void," or from *xáos*, a picture which supports the unresolvable/irreducible lawless hypothesis. The analysis presented in this paper suggests that the "laws" discovered in science correspond merely to syntactical correlations, are local and not universal.

Keywords: Number world; spurious law; emergent law

---

1. Introduction

What if the universe, on the most fundamental layer, just consisted of numbers? This is a suspicion at least as old as the Pythagoreans. According to Huffman's entry in *The Stanford Encyclopaedia of Philosophy* [1], "... in the *Metaphysics*, he [Aristotle] treats most Pythagoreans as adopting a mainstream system in contrast to another group of Pythagoreans whose system is based on the table of opposites .... The central thesis of the mainstream system is stated in two basic ways: the Pythagoreans say that things are numbers or that they are made out of numbers. In his most extended account of the system in *Metaphysics* 1.5, Aristotle says that the Pythagoreans were led to this view by noticing more similarities between things and numbers than between things and the elements, such as fire and water, adopted by earlier thinkers." Moreover, according to another contemporary review ([2], p. 14), "according to Aristotle, the Pythagoreans do not place the objects of mathematics between the ideas and material things as Plato does, they say 'that things themselves are numbers' and that 'number is the matter of things as well as the form of their modifications and permanent states'. As the principles of mathematics, numbers are the 'principles of all existing things'."

The importance of number, and more generally, of mathematics, for not only describing but "being" the bricks of the universe was stressed by eminent physicists, like Schrödinger ([3], Chapter III). The introduction of computing machinery creating virtual realities brought these issues to the forefront [4–9]. In a recent bold leap, Tegmark's *Mathematical Universe Hypothesis* [10,11] states that "the physical universe is not merely described by mathematics, but is a mathematical structure". As a consequence, mathematical existence equals physical existence, and all structures that exist mathematically (even in a non-constructive way) exist physically as well.

How could things be numbers? A world "spanned" by numbers can be represented by a single infinite (binary) sequence[1], or, equivalently, a single real number.

---

[1]   *A sequence is infinite while a string is finite. A finite prefix of a sequence is then a string.*

In what follows a *number world* will be modelled by a (binary) sequence.[2] Our choice is not to operate with the more geometric-centred Ancient Greek concept of number, which is essential for many continuous models of mathematical physics, but with an algorithmic one which is capable of giving a global perspective of the universe. Adopting this framework is motivated by Plato's mathematical discussion, in *Timaeus*, of the relations between numbers and things, see [2], p. 14 and also [12], and it is adopted here as a matter of *hypothesis*.

All entities encoded therein, including observers as well as measured objects, must be embedded in [6,13]; that is, they must themselves be (formed out of) numbers or symbols [14]. Non-numeric properties associated with such a "world on a sequence" can arise by way of a structural, levelled hierarchy [15].

Epistemologically this can be perceived as "emergence[3] of reality", which is the inverse of reductionism to some more fundamental, basic levels, involving explanations in terms of ever "smaller" entities: physical/universal/natural laws—in particular, relational and probabilistic ones—arise as effective patterns and structures "bottom-up" (rather than "top-down").

Such concepts were quite popular in the *fin de siécle* Viennese physical circles, so much so that they have been referred to as the *Austrian Revolt in Classical Mechanics* [18] and *Vienna Indeterminism* [19]: stimulated by the apparent indeterminacy manifesting in Rutherford's asymptotic decay law and its corroboration by Schweidler [20], Exner's 1908 inaugural lecture as *Rector Magnificus* included the suggestion that ([21], p. 18) "we have to perceive all so-called exact laws as probabilistic which are not valid with absolute certainty; but the more individual processes are involved the higher their certainty". Also Schrödinger's inaugural lecture in Zürich entitled *"What is a natural law?"* adopted and promoted Exner's ideas [22,23], well in accord with Born's later inclinations [24]. Since then classical statistical physics, as well as radioactive decay processes and quantised systems have operated under the presumption that the most fundamental layers of microphysical description are—both theoretically as well as phenomenologically and empirically—consistent with irreducible indeterminism.

Later related ideas have been brought forward in the context of a layered structure of physical theories [15], emergent cognition—perceived as an "emerging epiphenomenon" of neural activity; not unlike traffic jams they arise from the movements of individual taxis [25] –as well as emergent computation [26].

In what follows we shall, in a "Humean spirit" [27], study "laws" as patterns/correlations in sequences using the concept of *spurious correlations in data*, to be defined later. Two guiding theories will be applied: one is algorithmic information theory, the other is Ramsey theory. The gist of these two ways of looking at data is twofold: "all very long, even irregular" data sequences contain "very large" (indeed, as long as you prefer) regular, computable and thus, in physical terms, *deterministic*, subsequences. Secondly, it is impossible and inevitable for any arbitrary data set *not* to contain a variety of spurious correlations; that is, relational properties which could physically be wrongly interpreted as laws "governing" that universe of data.

2. Physical/Universal/Natural Laws

The notion of law in natural sciences, or law of the universe [28–33] has a long ambivalent history. It might not be overstated to claim that the conjecture that there are laws of nature is the core to what science is and how it was and is performed. Of course, one can refute this view and this lawless hypothesis has been discussed by various authors, see [12,34–40]. Contemplating a lawful universe usually amounts to assuming that the laws of nature are objective, have always existed and will exist,

---

[2] This physical-mathematical mapping assumption is essential for this paper.

[3] "Emergence is a notorious philosophical term of art." [16]. In this paper we will not use the term in the sense of the philosophical emergency theory, but with the signification given in physics [17]: *"The term emergent is used to evoke collective behaviour of a large number of microscopic constituents that is qualitatively different than the behaviours of the individual constituents."*

and they are written in the language of mathematics. Taken this for granted is an assumption which raises many problems, some of which will be discussed later. In this tradition science can be done in one way, the Galileo-Newton one; but if there are no laws, we can be freed to pursue other methodological options, some of which are not entirely unproblematic. Continuing to enrich the fundamental Greek practice of scientific observation, thinking and debating on different theoretical interpretations of phenomena with other methods, like the experimental methods (since Galileo) and the mathematical models (since Descartes and Newton) is obviously desirable. A step in this direction is to incorporate robust data analytics as a scientific method, see [41–44]. However, suggestions to narrow down the scientific methods to just a collection of "empirical evidences", to advance purely speculative theories (see [45] for physics) or to promote the "philosophy" according to which correlation supersedes causation and theorising (see [46]) are dangerous.[4]

The laws governing "physis" (nature) and those under which human societies are ruled have often been conflated and postulated to be of the same origin. At the dawn of western civilisation Heraclitus held that *lógos*[5] permeates everything, an arrangement common to all things yet incomprehensible to man ([48,49] and [50]). However, there are crucial differences between these laws. As Aristotle argued, a law is "by nature" if it is justified by appeal to something other than an agreement or a decision; in contrast, the laws human societies are ruled by are agreed upon in the Agora. While the former laws have been considered "absolute", the latter are clearly conventional. For example, the laws of movement are natural in contrast with the institutional structure of Greek democracy which is the result of human consensus. In Rhetoric, I.13, Aristotle discusses also the compatibility between the natural and the conventional laws. a characteristic of human justice, in contrast to divine justice. Both these laws are different from the concept of "natural law" developed in the Greek (Aristotle) as well as the Roman (Cicero) philosophies. In this philosophical sense a "natural law" asserts certain rights inherent by virtue of *human nature*. Endowed by nature—by God or a transcendent source—such a law can be understood universally through human reason [51]. Two typical laws of Aristotelian "physis" are: (i) Nothing moves unless one pushes it (there must be a 'mover' in order to move it). (ii) Because motion does exist, the above law implies that there must be a self-moved mover, i.e., a 'Prime Mover'. Finally, according to the definition of "natural" found in the Nicomachean Ethics, V.7, God is both a lawgiver for humans and the governor of nature, a view which was inherited by Christianity.

3. Laws and Limit Constructions

The scientific revolution grounded the proposal of new laws of nature on observation and iterable experimentation; sometimes these types of laws were simply guessed or invented, but nonetheless on the grounds of a "meaningful" (physical, theoretical and practical) framework. For example, after several experiments, some of which were just imagined, Galileo and others [52] proposed the "law" of inertia. This law is a fundamental conservation principle, the conservation of momentum, and is a limit principle since no physical body actually moves at constant speed along an Euclidean line–a straight line with no thickness. Yet, by extrapolating from his observations made on the object of bodies as their friction was changed, Galileo was able to deduce the concept of inertia, and closely analyse what circumstances affect this asymptotic movement: friction and gravitation. Thus, by this scientific process of induction, deduction, extrapolation and abduction [53–55], an Aristotelian, God given, absolute, notion of a law of "physis" was radically modified. The advantage of this notion of physical law based on limit principles and symmetry is visible once Newton made the connection between falling apples and planets: there is no need to be anyone pulling nor pushing the planets to move them around. Indeed, Newton's law of gravitation gives the trajectories of any two bodies in inertial movement within a gravitational field, including apples and planets. On the one hand it became possible to derive

---

[4]   See the Appendix A for a more formal discussion.
[5]   Lógos is the apparent antithesis of *xáos* in Hesiod's *Theogony* [47].

Kepler's trajectories and laws for one sun and one planet from Newton's law, without the need for a Prime Mover that is constantly pushing. On the other hand, Newton realised that, with two or more planets, reciprocal interactions destabilise the planets' trajectories (which later would be recognised as a result of chaotic non-linearity). He thus assumed the aid of occasional interventions of God in order to assure the stability of the planetary system *in secula seculorum*: God, through a few sapient touches, was the only guarantor of the long term stability of the Solar System [56]. Poincaré later confirmed mathematically this deep intuition of Newton on the asymptotic chaos within the Solar System (see below for more discussion of this). We should note, however, that this analysis only makes sense in the mathematical continua. Inertia is conceived as a limit property; moreover, its understanding as a conservation law (of momentum) alongside the conservation of energy, as a symmetry in the equations (as a result of Noether's theorems relating symmetries to conserved quantities [57]), is based on continuous symmetries: they are invariant with respect to continuous translations in space or time. A few years later, Galileo, Boyle and Mariotte proposed another limit law: they traced the isothermal hyperbolas of pressure and volume for perfect gases. Of course, actual gases, as a result of friction, gravitation, inter-particle interactions, etc., do not follow this peculiar conic section; yet its abstract, algebraicma formulation and its geometric representation, allowed a uniform and general understanding of the earliest law of thermodynamics. Principles referring to inexistent ideal trajectories, at the external limit of phenomena, continued to rule knowledge constructions in physics. As another example, let us consider Boltzmann's ergodic principle: *In a perfect gas a particle stays in a region of a given space for an amount of time proportional to the volume of that region.* Once again this is an asymptotic principle, as it uniformly holds only at the infinite limit in time. On these grounds, Boltzmann's thermodynamic integral that allows the deduction of the second law of thermodynamics (regarding the increase in entropy) is also formulated as a limit construction (an integral): it holds only at the infinite limit of the number of particles in the volume of gas. Can one prove, or at least corroborate, these asymptotic principles? There is no way to put oneself or a measurement instrument at these limit conditions and check for Euclidean straight lines, hyperbolas or behaviour at the asymptotic limit in time. One may only try to falsify some consequences [58]; yet, even in such cases the derivation itself may be wrong, but not necessarily the principle. As has already been pointed out by many philosophers, among them Hume, Berkeley, Kant and Schopenhauer, all we can produce—and this is a crucial point—is *scientific knowledge*: we understand a lot, but not everything, through these limit principles that unify all movements, all gases, etc., as specific instances of inexistent movements and gases. And, more importantly, as a result we can construct fantastic tools and machines that work reasonably well – but not perfectly well, of course—and have radically changed our lives. With these machines the westerners dominated the world after the scientific revolution, a non trivial consequence of their science and its "absolute" laws. We are typing, reading and exchanging data in networks of the latest of these inventions, an excellent, but not perfect, instance of a limit machine–the Turing machine. One of the limit principles of these machines is Turing's distinction between hardware and software and the identification between program with data that allows abstract, mathematical styles of programming all the while (almost) disregarding their material realisation.

Another important consequence was the discovery of limits of computing, specifically the incomputability of the halting problem, and more generally the development of theoretical computer science [59].[6] At the same time the abstract character obscured the role played by physics in computing: because of the separation between hardware and software, the role of hardware in computation was largely ignored in theoretical computer science, arguably delaying with a few decades the understanding and development of physics of computation, reversible computing and quantum computing, [61–63].

---

[6] These limits can be mitigated from a practical point of view with various methods; for example, the halting problem can be solved probabilistically with arbitrarily high precision [60].

4. Order within Disordered Sequences

In intuitive terms, Ramsey theory states that there exists a certain degree of order in all sets/sequences/strings, regardless of their composition. Heuristically speaking, this is so because it is impossible for a collection of data not to have "spurious" correlations, that is, relational properties among its constituents which are determined only by the size of the data. The simplest example of such (spurious) correlation is given by the *Dirichlet's pigeonhole principle* stating that n pigeons sitting in $m < n$ holes result in at least one hole being filled with at least two pigeons. Or in a party of any six people, some three of them are either mutually acquaintances, or complete strangers to each other [64,65].[7] This seemingly obvious statements can be used to demonstrate unexpected results; for example, the pigeonhole principle implies that there are two people in Paris who have the same number of hairs on their heads. The pigeonhole principle is true for at least two pigeons and one whole; the party result needs at least six people. A common drawback of both results is their non-effectivity: we know that two people in Paris have the same number of hairs on their heads, but we don't know who they are.

An important result in Ramsey theory is Van der Waerden theorem (see [66]) which states that *in every binary sequence at least one of the two symbols must occur in arithmetical progressions of every length.*[8] The theorem describes a set of arbitrary large strong correlations – in the sequence $x_1 x_2 \ldots x_n \ldots$ there exist arbitrary large k, N such that equidistant positions $k, k + t, k + 2t, \ldots k + Nt$ contain the same element (0 or 1), that is, $x_k = x_{k+t} = x_{k+2t}, \cdots = x_{k+Nt}$.[9] Crucial here is the fact that the property holds true for *every* sequence, ordered or disordered.[10] Are these correlations "spurious"? According to Oxford Dictionary, *spurious* means "Not being what it purports to be; false or fake. False, although seeming to be genuine. Based on false ideas or ways of thinking." The (dictionary) definition of the word "spurious" is semantic, that is, it depends on an assumed theory: one correlation can be spurious according to one theory, but meaningful with respect to another one.

Can we give a definition of "spurious correlation" which is independent of *any* theory? Following [46] a *spurious correlation* is defined in a very restrictive way as follows: *a correlation is spurious if it appears in a randomly generated string/sequence.* Indeed, in the above sense a spurious correlation is "meaningless" according to any reasonable interpretation because, by construction, its values have been generated at "random", as all data in the sequence. As a consequence, such a correlation cannot provide reliable information on future developments of any type of behaviour. Of course, there are other reasons making a correlation spurious, even within a "non-random" string/sequence. But, are there correlations as defined above? Van der Waerden theorem proves that in every sequence there are spurious correlations in the above sense – they can be said to "emerge". Therefore, these spurious correlations can also be re-interpreted as "emerging laws." It is important to keep in mind that these "laws" are not properties of a particular sequence,—indeed, they exist in *all* sequences as Van der Waerden theorem proves. How do the spurious correlations manifest themselves in a number world? From the finite version of Van der Waerden theorem, the more bits of the sequence describing the number world we can observe, the longer are the lengths of monochromatic arithmetical progressions. So, once there are (sufficiently many) data, regardless of their intrinsic structure, "laws from nowhere" (*ex nihilo*) emerge. In what follows we will work only with the above definition of spurious correlation.

Are these spurious correlations just simple accidents or more customary phenomena? We can answer this question by analysing the "sizes" of the sets of random sequences/strings in which spurious correlations arise. As our definition of spurious correlation is independent of any theory, in

---

[7]   In fact, there is a second trio who are either mutually acquainted or unacquainted [64].

[8]   If we interpret 0 and 1 as colours, then the theorem says that in every binary sequence there exist arbitrarily long monochromatic arithmetical progressions.

[9]   Again, the proof is not constructive.

[10]   The finite version of Van der Waerden theorem shows that the same phenomenon appears in long enough strings. See more in [46].

answering the above questions we will use a model of randomness for sequences and strings provided by algorithmic information theory [67,68] which has the same property.

First, how "large" is the set of random sequences? If we work with Martin-Löf random sequences[11], then the answer is "almost all sequences": the probability of a sequence to be Martin-Löf random is one.[12] This means that *the probability that an arbitrary sequence does not have spurious correlations is zero.*[13]

Second, as human access to sequences is limited to their finite prefixes, it is necessary to answer the same question for strings: what is the "size" of "random" strings? Using the incompressibility criterion again [46], a string x of length n is α-random if no Turing machine can produce x from an input with less than $n - \alpha \cdot n$ bits.[14] The number of α-random strings x of length n is larger than $2^n \left(1 - 2^{-\alpha \cdot n}\right) + 1$, and hence, with finitely many exceptions, it outnumbers the number of binary strings of length n which are not α-random.[15] More interestingly, the probability that a string x of length n is α-random is larger than $1 - 2^{-\alpha \cdot n} + 2^{-n}$, an expression which tends exponentially to 1 as n tends to infinity. This means that *the probability that an arbitrary string does not have spurious correlations is as close to zero as we wish provided that its length is large enough, that is, excluding finitely many strings.*

Furthermore, the increase of some types of spurious correlations, i.e., emergent "laws", can be quantified: Goodman's inequality [69,70] yields lower bounds on how many spurious correlations are observed as a function of the size of data. Conversely, Pawliuk recently suggested [71] that Goodman's inequality can be utilised for testing the (null) hypothesis that a dataset is random: if the bounds are over-satisfied, the correlations might be not spurious, and thus the dataset might not be stochastic. Can we distinguish between meaningful laws and emerging "laws"? The answer seems to be negative at least from a computational point of view.

5. The Emergence of Turing Complete (Universal) Computation

In view of the "quantification" of information content [67,72], how could complexity and structures such as universal computation, evolve even in principle? The answer to this question is in the algorithmic information content (complexity) of the number world.

The proof of Turing completeness[16] of the Game of Life provided by Conway in ([73], Chapter 25, What Is Life?) is a useful method for exploring how complex behaviour like Turing completeness can emerge from very simple rules, in this case, the rules of cellular automata (see more in [74]). With a universal Turing machine and all α-random strings one can generate *all* strings [67].

Is this phenomenon also possible for sequences, that is, for number worlds? The answer is affirmative. According to a theorem by Kučera-Gács-Hertlinger ([67], p. 179), there effectively exists a process F—which is continuous computable operator—which generates all sequences from the set of Martin-Löf random sequences: in other words, every sequence is the image from F of a Martin-Löf random sequence.

6. Is the World Number Computable?

Of course, there exist infinitely (countable) computable world numbers.

---

[11] A Turing machine with a prefix-free domain is called self-delimiting. A (self-delimiting) Turing machine which can simulate any other (self-delimiting) Turing machine is called universal. A sequence is Martin-Löf random if there exists a fixed constant such that every finite prefix (string) of the sequence cannot be compressed by a self-delimiting universal Turing machine by more than a constant [67].

[12] This holds true even constructively.

[13] Probability zero is not the same as impossibility: there exist infinitely many sequences—like the computable ones—which contain no spurious correlations.

[14] The minimum length of an input a Turing machine needs to compute a string of length n lies in the interval $(0, n + c)$, where c is a fixed constant. From this it follows that $\alpha \in (0, 1)$.

[15] More precisely, when $n \geq 2/\alpha$.

[16] A model of computation is Turing complete—sometimes called universal—if it can simulate a universal Turing machine.

Can we decide whether the sequence describing a given world number is computable? Answering this question is probably impossible both theoretically and empirically. However, we can answer a simpler variant of the question: What is the probability that a world number is computable? If we take as probability the Lebesgue measure [67], then the answer is zero.[17]

The above result shows that the probability that a world number can be generated by an algorithm is zero. If we weaken the above requirement and ask about the probability that there exists an algorithm which generates infinitely many bits of a world number, then the answer remains the same: this probability is nil. This result follows from a theorem in algorithmic information theory saying that the complement of the above set—the set of bi-immune sequences[18] —has probability one [67]. A consequence of this fact, corroborated by an extension of the Kochen-Specker theorem proving value indefiniteness of quantum observables relative to rather weak physical assumptions [75], is that with probability one a number world is produced by repeatedly measuring of such a value indefinite observable.

7. Non-Uniform Evolution

Two examples of world numbers are particularly interesting: Champernowne world number and Chaitin world number. A Champernowne world number in base two is given by the sequence

$$C_2 = 0100011011000001010011100101110111010000\ldots$$

which consists in the concatenation of all binary strings enumerated in quasi-lexicographical order [76].[19] A a Chaitin world number is given by a Chaitin Ω_U number (or halting probability), that is the probability that the universal self-delimiting Turing machine U halts [77]. Chaitin world numbers "hold proofs" for almost all mathematical known results; such as as Fermat Last Theorem (in the 400 initial bits), Goldbach's conjecture, or important conjectures like Riemann Hypothesis (in the 2745 bits initial bits) and P vs. NP (in the 6,495 initial bits; cf. [78]. Both world numbers are Borel normal in the sense that every binary string x appears in these sequences infinitely many times with the same frequency, namely $2^{-|x|}$, where $|x|$ is the length of x. In such a world every text—codified in binary—which was written and will be ever written appears infinitely many times and with the same frequency, which depends only on the length of the text. In particular, any correlation appears in such a world infinitely many times. However, these worlds are also very different: A Champernowne world number is computable, but a Chaitin world number is highly incomputable because it is Martin-Löf random. As a consequence, while both number worlds have all possible correlations repeated infinitely many times, the status of those correlations are different: in a Chaitin world number these correlations are spurious (because of its randomness), but in a Champernowne world number they are not (because its computability, hence highly non-randomness).

How an embedded observer would "feel" to live in such a world? This is a deep question which needs more study. Here we will make only a few simple remarks (see also [36]).

First, no observer or rational agent could decide in a finite time whether they live in a Champernowne or Chaitin world. Second, any observer or rational agent surviving, or at least recording experimental outcomes, a sufficiently long time will see many of the previously discovered accepted "laws" being refuted.

Third, suppose intrinsic observers embedded into a mathematical universe experience and "surf" these number worlds by their interactions with them; that is, they perceive long successions of initial

---

¹⁷ Again, one should not think that this means that there are no computable world numbers, see Section 6. The result follows from the fact that the computable sequences form a countable set.

¹⁸ A sequence is bi-immune if its corresponding set of natural numbers nor its complement contain an infinite computably enumerable subset.

¹⁹ In base 10, $C_{10} = 12345678910111213141516\ldots$.

bits of their defining infinite sequences. Assume now that these sequences are Champernowne or Chaitin sequences. Because of the Borel normality of these sequences, the strings surfed by observers are Borel normal as finite objects, that is, they are distributed uniformly up to finite corrections [79]. How would intrinsic observers experience such variations? In one scenario one may speculate that intrinsically such "interim" periods of monotony may not count at all; that is, these will not be operationally recognised as such: for an embedded observer [6,13], the world number will remain "dormant" while the number world remains monotonous.

Another option, maybe even more speculative, is to assume that, as long as the world number allows for a sufficiently wide variety of substrings, the intrinsic phenomenology will, through emerging character of (self-) perception, "pick" its own segment or pieces (of numbers) from all the available ones. Indeed, it might not matter at all for intrinsic perception whether, for instance, the cycle time is altered (reduced, increased), or whether the lapse of cycles is arbitrarily exchanged or even inverted: as long as there are still "sufficient" patterns and number states emerging could "process" and "use," lawfulness and consciousness will always ensue [80].

8. Is the Universe Lawless?

In this section we add another argument—to many others [12,34–40]—in favour of the hypothesis in the title.

There are uncountable infinite binary sequences, each of which could be a (the) "true" simplest model of our universe. Among these candidates, we have the set of Martin-Löf random sequences, which will fit with the hypothesis: this set is very large, because as we have already mentioned, the probability that an infinite binary sequence is not Martin-Löf random is (constructively) zero. However, the complement of this set—which has then probability zero—is not only infinite, but also uncountable and therefore cannot be lightly discarded.

The so-called physical/universal/natural "laws" deal with the infinity, on one hand; but can be verified only on finitely many cases. What about the situation when a "true model" is not a Martin-Löf random sequence, possibly a highly improbable computable one?

In order to be able to attempt to confirm the "laws" in this model we have to surf the initial bits of the infinite sequence. How many bits can be surfed? A possible bound from below is the number of atoms in the universe which is believed to be less than $\text{Number}_{atoms} = 10^{82}$. What is then the probability that an infinite sequence, thought as a model of our universe, starts with an α-random string of length Number_{atoms}? In this set there are infinitely many Martin-Löf random sequences and a sequence is Martin-Löf random with probability one, see Section 6, but also infinitely many computable sequences. The analysis in Section 4 shows that this probability will be larger than

$$1 - 2^{-\alpha \cdot 10^{82}} + 2^{-10^{82}},$$

because this is the probability that an infinite binary sequence starts with an α-random string of length Number_{atoms}. With this probability—which is infinitesimally close to one—*every choice for our model of our universe* starts with an α-random string; consequently, all patterns and correlations which can be verified in this model are spurious!

Let us hasten to note that spurious does not mean wrong, not genuine, useless. On the contrary, correlations can be, and many times are, interesting, useful and give us insight about the working of the universe; they are, however, local and not universal.

9. Conclusions

According to Heidegger [81], the most profound and foundational metaphysical issue is to think the existent as the existent (*"das Seiende als das Seiende denken"*). Here the existent is metaphorically interpreted as an infinite sequence of bits, a Number World. Rather than answering the primary question [82] of why there is existence rather than nothingness, this paper has been mostly concerned with the formal consequences of existence under the least amount of extra assumptions [83].

As it turns out, existence implies that an intrinsic and sophisticated mixture of meaningful and (spurious) patterns—possibly interpreted as "laws"—can arise from xáos. The emergent "laws" abound, they can be found almost everywhere. The axioms in mathematics find their correspondents in the "laws" of physics as a sort of "lógos" upon which the respective mathematical universe is "created by the formal system". By analogy, our own universe might be, possibly deceptively and hallucinatory, be perceived as based upon such sorts of "laws" of physics. The results in Sections 4 and 8 have corroborated the Humeanism view, later promoted by Exner and to some extend by the young Schrödinger, that at least some physical "laws" merely arise from xáos; a picture which is compatible with the unresolvable/irreducible lawless hypothesis. The analysis presented in this paper suggests that the "laws" discovered in science correspond merely to syntactical correlations, are local and not universal.

As in biological living systems, the dynamics described above is not a matter of stable or unstable equilibrium, but of far from equilibrium processes which are "structurally stable". This "duality" is supported in physics by the hierarchical layers theory [15,84]. The simultaneous structural stability and non-conservative behaviour in biology, which is a blend of stability and instability is due to the coexistence of opposite properties such as order/disorder and integration/differentiation [85].

Such an active and mindful (some might say self-delusional and projective) approach to order in and purpose of the universe may be interpreted in accord with the ancient Greek theogony [47] by saying that *lógos*, the Gods and the laws of the universe, originate from "the void," or, in a less certain interpretation, from xáos. Very similar concepts were developed in ancient China probably around the same time as Homerus and Hesiod: the *I Ching* utilises relational properties of symbols from sophisticated stochastic procedures providing inspiration, meaning and advice on what has been understood as divine intent and the way the universe operates.

Author Contributions: Conceptualization, C.S.C. and K.S.; methodology, C.S.C. and K.S.; writing—original draft preparation, C.S.C.; writing—review and editing, C.S.C. and K.S.

Funding: K. Svozil was supported in part by the John Templeton Foundation's *Randomness and Providence: An Abrahamic Inquiry Project*

Acknowledgments: We thank the anonymous referees for useful comments, criticism and suggestions which have significantly improved the paper

Conflicts of Interest: The authors declare no conflict of interest.

Appendix A. Causation and Correlation: Two Formal Models

To understand better the difference between causation and correlation we present two simple models. In the first model we have two hypotheses, x and y which can true or false and we denote by $x \succ y$ the proposition "x is a cause for y" and by $C(x,y)$ the proposition "x and y are correlated". The logical representations of the new propositions are enumerated in the following Table A1: Indeed, $x \succ y = 1$ if x is true, then y is true, that is, $x = y = 1$. Note that $x \succ y$ is a "more restrictive" operator than the logical implication which is true also when $0 \to y = 1$, for every $y \in \{0,1\}$. We have $C(x, y = 1)$ if and only if both x and y are either true or false, that is, $x = y$. If follows that $x \succ y$ implies $C(x,y) = 1$, but the converse is false.

Table A1. Causation versus correlation: a logical model.

x	y	$x \succ y$	$C(x, y)$
0	0	0	1
0	1	0	0
1	0	0	0
1	1	1	1

The second model is inspired by the Fechner-Machian identification of causation with functional dependence [32,86]: suppose that data is represented by two sets X and Y. If $f\colon X \to Y$ is a function from X to Y, then we denote the graph of f by $G_f = \{(x, f(x)) \in X \times Y \mid f(x) = y\}$. A relation R between X and Y is a set $R \subseteq X \times Y$. We say that $x \in X$ is an f-cause for $y \in Y$ if $f(x) = y$ and we write $x \succ_f y$. The elements x, y are correlated by the relation R, in writing, $C_R(x,y)$, if $(x,y) \in R$. Assume that $G_f \subset R$; if $x \succ_f y$, then $C_R(x,y)$ but the converse implication is not true.

Both models show that correlation is symmetric, but causation is not. However, the models above do not reflect a crucial difference between causation and correlation: the former contributes to the understanding, in an imperfect way, of the phenomenon, but the second is just a syntactical observation. Causation invites testing, revision, even abandonment; correlation is static and without further analysis could be misleading, see [87,88].

References

1. Huffman, C. Pythagoreanism. In *The Stanford Encyclopedia of Philosophy*; Zalta, E.N., Ed.; Metaphysics Research Lab Stanford University: Stanford, CA, USA, 2016.
2. Maziarz, E.A.; Greenwood, T. *Greek Mathematical Philosophy*; Ungar: New York, NY, USA, 1968.
3. Schrödinger, E. *Nature and the Greeks*; Cambridge University Press: Cambridge, UK, 1954, 2014.
4. Fredkin, E. Digital mechanics. An informational process based on reversible universal cellular automata. *Physica* **1990**, *D45*, 254–270.
5. Margolus, N. Physics-like model of computation. *Physica* **1984**, *D10*, 81–95.
6. Toffoli, T. The role of the observer in uniform systems. In *Applied General Systems Research: Recent Developments and Trends*; Klir, G.J., Ed.; Plenum Press, Springer: New York, NY, USA; London, UK, 1978; pp. 395–400.
7. Wolfram, S. *A New Kind of Science*; Wolfram Media, Inc.: Champaign, IL, USA, 2002.
8. Zuse, K. *Rechnender Raum*; Friedrich Vieweg & Sohn: Braunschweig, Germany, 1969.
9. Zuse, K. *Calculating Space*; MIT Technical Translation AZT-70-164-GEMIT; MIT (Proj. MAC): Cambridge, MA, USA, 1970.
10. Tegmark, M. The mathematical universe. *Found. Phys.* **2007**, *38*, 101–150. [CrossRef]
11. Tegmark, M. *Our Mathematical Universe: My Quest for the Ultimate Nature of Reality*; Penguin Random House LLC.: New York, NY, USA, 2014.
12. Calude, C.S.; Meyerstein, W.F.; Salomaa, A. The Universe is Lawless. In *Computable Universe: Understanding and Exploring Nature as Computation*; World Scientific: Singapore, 2013; pp. 527–537.
13. Svozil, K. Extrinsic-intrinsic concept and complementarity. In *Inside versus Outside*; Atmanspacher, H., Dalenoort, G.J., Eds.; Springer Series in Synergetics: Berlin, Germany, 1994; pp. 273–288.
14. Borges, J.L. *La biblioteca de Babel—The Library of Babel*; Editorial Sur: Buenos Aires, Argentina, 1941; (in Spanish), 1962 (in English). See also Borges' 1939 essay "The Total Library".
15. Anderson, P.W. More is different. Broken symmetry and the nature of the hierarchical structure of science. *Science* **1972**, *177*, 393–396. [CrossRef] [PubMed]
16. O'Connor, T.; Wong, H.Y. Emergent properties. In *The Stanford Encyclopedia of Philosophy*; Zalta, E.N., Ed.; Metaphysics Research Lab Stanford University: Stanford, CA, USA, 2015.
17. Kivelson, S.; Kivelson, S.A. Defining emergence in physics. *Npj Quant. Mater.* **2016**, *1*, 16024. [CrossRef]
18. Hiebert, E.N. Common frontiers of the exact sciences and the humanities. *Phys. Perspect.* **2000**, *2*, 6–29. [CrossRef]

19. Stöltzner, M. Vienna indeterminism: Mach, Boltzmann, Exner. *Synthese* **1999**, *119*, 85–111. [CrossRef]
20. Schweidler, E.v. *Über Schwankungen der radioaktiven Umwandlung*; H. Dunod & E. Pinat: Paris, France, 1906; pages German part, 1–3.
21. Exner, F.S. *Über Gesetze in Naturwissenschaft und Humanistik: Inaugurationsrede gehalten am 15. Oktober 1908*; Hölder, Ebooks on Demand Universitätsbibliothek: Viena, Austria, 1909.
22. Schrödinger, E. Was ist ein Naturgesetz? *Naturwissenschaften* **1929**, *17*, 01. [CrossRef]
23. Schrödinger, E. *Science And The Human Temperament*; George Allen & Unwin: Crows Nest, UK, 1935.
24. Born, M. Zur Quantenmechanik der Stoßvorgänge. *Zeitschrift Phys.* **1926**, *37*, 863–867. [CrossRef]
25. Hofstadter, D.R. *Artificial Intelligence: Subcognition as Xomputation*; Technical Report TR132; Indiana University: Bloomington, IN, USA, 1982.
26. Forrest, S. Emergent computation: Self-organizing, collective, and cooperative phenomena in natural and artificial computing networks: Introduction to the proceedings of the ninth annual cnls conference. *Phys. D Nonlinear Phenomena* **1990**, *42*, 1–11. [CrossRef]
27. Hume, D. The Clarendon Edition of the Works of David Hume. In *A Treatise of Human Nature: Volume 1: Texts*; Clarendon Press and Oxford University Press: Oxford, UK, 2007;
28. Beebee, P. Hume and the problem of causation. In *The Oxford Handbook of Hume*; Russell, P., Ed.; Oxford Handbooks; Oxford University Press: Oxford, UK; New York, NY, USA, 2016.
29. De Pierris, G.; Friedman, M. Kant and Hume on causality. In *The Stanford Encyclopedia of Philosophy*; Zalta, E.N., Ed.; Metaphysics Research Lab, Stanford University: Stanford, CA, USA, 2013.
30. Feynman, R.P. *The Character of Physical Law*; MIT Press: Cambridge, MA, USA, 1965.
31. Mumford, S.; Anjum, R.L. *Causation: A Very Short Introduction*; Oxford University Press: Oxford, UK, 2014.
32. Norton, J.D. Causation as folk science. *Philos. Imprint* **2003**, *3*, 1–22.
33. Russell, B. I.–On the notion of cause. *Proc. Aristotelian Soc.* **1913**, *13*, 06. [CrossRef]
34. Armstrong, D.M. What is a Law of Nature? In *Cambridge Studies in Philosophy*; Cambridge University Press: Cambridge, UK, 1983.
35. Cabello, A. Physical origin of quantum nonlocality and contextuality. *arXiv* **2019**, arXiv:1801.06347.
36. Calude, C.S.; Meyerstein, W.F. Is the universe lawful? *Chaos Solit. Fract.* **1999**, *10*, 1075–1084.
37. Mueller, M.P. Could the physical world be emergent instead of fundamental, and why should we ask? (short version). *arXiv* **2017**, arXiv:1712.01826.
38. Rosen, J. *Lawless Universe: Science and the Hunt for Reality*; The John Hopkins University Press: Baltimore, MA, USA, 2010.
39. van Fraassen, B.C. *Laws and Symmetry*; Oxford University Press: Oxford, UK, 1989, 2003.
40. Yanofsky, N. Chaos makes the multiverse unnecessary. *Nautilus* **2017**, *66*, 1–16.
41. Rajaraman, A.; Ullman, J.D. *Mining of Massive Datasets*; Cambridge University Press: Cambridge, UK, 2011.
42. Reed, D.A.; Dongarra, J. Exascale computing and big data. *Commun. ACM* **2015**, *58*, 56–68. [CrossRef]
43. Schutt, R.; O'Neil, C. *Doing Data Science*; O'Reilly Media: Newton, MA, USA, 2014.
44. Stanton, J.M. *Introduction to Data Science*. Syracuse University: Syracuse, NY, USA, 2012.
45. Ellis, G.; Silk, J. Scientific method: Defend the integrity of physics. *Nature* **2014**, *516*, 321–323. [CrossRef]
46. Calude, C.S.; Longo, G. The deluge of spurious correlations in big data. *Found. Sci.* **2017**, *22*, 595–612. [CrossRef]
47. Hesiod. *Hesiod: Volume I, Theogony. Works and Days. Testimonia (Loeb Classical Library No. 57)*; Harvard University Press: Cambridge, MA, USA; London, UK, 2006.
48. Curd, P.; McKirahan, R.D. *A Presocratics Reader (Second Edition). Selected Fragments and Testimonia*; Hackett Publishing Co.: Indianapolis, IN, USA; Cambridge, UK, 2011.
49. Diels, H.; Kranz, W. *Die Fragmente der Vorsokratiker*, 6th ed.; Weidmannsche Buchhandlung: Berlin, Germany, 1952.
50. Kirk, G.S.; Raven, J.E.; Schofield, M. *The Presocratic Philosophers: A Critical History with a Selcetion of Texts*, 2nd ed.; Cambridge University Press: Cambridge, UK, 1983.
51. Corbett, R.J. The question of natural law in Aristotle. *History Political Thought* **2009**, *30*, 229–250.
52. Drake, S. Galileo and the law of inertia. *Am. J. Phys.* **1964**, *32*, 601–608. [CrossRef]
53. Douven, I. Abduction. In *The Stanford Encyclopedia of Philosophy*; Zalta, E.N., Ed.; Metaphysics Research Lab Stanford University: Stanford, CA, USA, 2017.

54. Josephson, J.R.; Josephson, S.G. *Abductive Inference: Computation, Philosophy, Technology*; Cambridge University Press: Cambridge, UK; New York, NY, USA; Melbourne, Australia, 1994.

55. Peirce, C.S.; Hartshorne,C.; Weiss, P.; Burks, A.W. *Collected Papers of Charles Sanders Peirce*; Harvard University Press Belknap Press: Cambridge, MA, USA, 1932.

56. Laskar, J. A numerical experiment on the chaotic behaviour of the solar system. *Nature* **1989**, *338*, 237–238. [CrossRef]

57. Noether, E. Invariante Variationsprobleme. *Nachrichten von der Gesellschaft der Wissenschaften zu Göttingen, Mathematisch-Physikalische Klasse* **1918**, *1918*, 235–257.

58. Popper, K.R. *The Logic of Scientific Discovery*; Hutchinson & Co and Routledge: New York, NY, USA, 1959, 1992.

59. Gruska, J. *Foundations of Computing*; International Thompson Computer Press: London, UK, 1997.

60. Calude, C.S.; Dumitrescu, M. A probabilistic anytime algorithm for the halting problem. *Computability* **2018**, *7*, 259–271. [CrossRef]

61. Frank, M.; Vieri, C.; Ammer, J.; Love, N.; Margolus, N.H.; Knight, T. A scalable reversible computer in silicon. In *Unconventional Models of Computation*; Springer: Singapore, 1998; pp 183–200.

62. Landauer, R. Computation: A fundamental physical view. *Phys. Scr.* **1987**, *35*, 88. [CrossRef]

63. Mermin, D.N. *Quantum Computer Science*; Cambridge University Press: Cambridge, UK, 2007.

64. Bostwick, C.W.; Rainwater, J.; Baum, J.D. E1321. *Am. Math. Mon.* **1959**, *66*, 141–142. [CrossRef]

65. Greenwood, R.E.; Gleason, A.M. Combinatorial relations and chromatic graphs. *Can. J. Math.* **1955**, *7*, 1–7. [CrossRef]

66. Graham, R. Some of my favorite problems in Ramsey Theory. *INTEGERS Electronic J. Combinat. Number Theory* **2007**, *7*, #A2.

67. Calude, C. *Information and Randomness—An Algorithmic Perspective*, 2nd ed.; Springer: Berlin, Germany, 2002.

68. Downey, R.G.; Hirschfeldt, D.R. Algorithmic Randomness and Complexity. In *Theory and Applications of Computability*; Springer: Berlin, Germany, 2010.

69. Goodman, A.W. On sets of acquaintances and strangers at any party. *Am. Math. Mon.* **1959**, *66*, 11. [CrossRef]

70. Schwenk, A.J. Acquaintance graph party problem. *Am. Math. Mon.* **1972**, *79*, 12. [CrossRef]

71. Pawliuk, M.; Waddell, M.A. Using Ramsey Theory to Measure Unavoidable Spurious Correlations in Big Data. *Axioms* **2019**, *8*, 29. [CrossRef]

72. Chaitin, G.J. *Algorithmic Information Theory (Cambridge Tracts in Theoretical Computer Science)*, revised ed.; Cambridge University Press: Cambridge, UK, 2003.

73. Berlekamp, E.; Conway, J.H.; Guy, R. *Winning Ways*; Academic Press: New York, NY, USA, 1982.

74. Rendell, P. *Turing Machine Universality of the Game of Life*; Springer International Publishing: Cham, Switzerland, 2016.

75. Abbott, A.A.; Calude, C.S.; Svozil K. Value indefiniteness is almost everywhere. *Phys. Rev. A* **2014**, *89*, 032109–032116. [CrossRef]

76. Champernowne, D.G. The construction of decimals normal in the scale of ten. *J. Lond. Math. Soc.* **1933**, *8*, 254–260. [CrossRef]

77. Calude, C.S.; Chaitin, G.J. What is … a halting probability? *Notices AMS* **2007**, *57*, 236–237.

78. Calude, C.S.; Calude, E. The complexity of mathematical problems: An overview of results and open problems. *Int. J. Unconv. Comput.* **2013**, *9*, 327–343.

79. Calude, C. Borel normality and algorithmic randomness. In *Developments in Language Theory*; Rozenberg, G., Salomaa, A., Eds.; World Scientific: Singapore, 1994; pp. 113–129.

80. Egan, G. *Permutation City*; Millennium Orion Publishing Group: London, UK, 1994.

81. Heidegger, M. *Was ist Metaphysik?* Klostermann: Frankfurt, Germany, 1929.

82. Heidegger, M. Einführung in die Metaphysik (Freiburger Vorlesung Sommersemester 1935). In *Martin Heidegger Gesamtausgabe*. Klostermann: Frankfurt, Germany, 1935.

83. Jaynes, E.T. *Probability Theory: The Logic Of Science*; Cambridge University Press: Cambridge, UK, 2003.

84. Pattee, H.H. *Postscript: Unsolved Problems and Potential Applications of Hierarchy Theory*, Springer: Dordrecht, The Netherlands, 2012; pp. 111–124.

85. Buiatti, M.; Longo, G. Randomness and multilevel interactions in biology. *Theory Biosci.* **2013**, *132*, 139–158. [CrossRef] [PubMed]

86. Heidelberger, M. Functional relations and causality in Fechner and Mach. *Philos. Psychol.* **2010**, *04*, 201023. [CrossRef]
87. Spurious Correlations. Available online: http://www.tylervigen.com/spurious-correlations (accessed on 9 April 2019).
88. Vigen, T. *Spurious Correlations*; Hachette Books: New York, NY, USA, 2015.

 *philosophies*

Article

What Is Physical Information?

Roman Krzanowski

Department of Philosophy, The Pontifical University of John Paul II, 31-002 Krakow, Poland;
rmkrzan@gmail.com

Received: 27 April 2020; Accepted: 18 June 2020; Published: 24 June 2020

Abstract: This paper presents the concept of physical information, and it discusses what physical information is, and how it can be defined. The existence of physical information has been discussed in several studies which recognize that properties of information are characteristic of physical phenomena. That is, information has an objective existence, a lack of meaning, and can be quantified. In addition, these studies recognize how a phenomenon that is denoted as physical information can be expressed as an organization of natural or artificial entities. This paper argues that concepts of (abstract) information that are associated with meaning also depend (to a substantial degree) on physical information.

Keywords: physical information; abstract information; physical phenomena; abstract entities

1. Introduction

In this paper, we study information that is defined as physical concrete phenomena. We also discuss its properties, and its relation to abstract concepts of information. This section presents the background to this study, the justification for undertaking it, and its potential benefits. This section also introduces some basic terms used within the paper.

The various definitions of information range from abstract notions of knowledge, messages, symbols, and signals [1–7], through to more physical notions as objective phenomena [8–11], and on to more quantified concepts—often expressing information as a series of bits [5,11–16].[1] It seems, however, that most concepts of information can be subsumed under two broad categories.[2]

In the first category, we bring together concepts that present information as something abstract, while in the second category, we can gather concepts that define information as a concrete physical phenomenon.

Information in the first category is presented as an abstract concept, therefore, we denote it as information$_A$, where the "A" represents "abstract".[3] In some definitions, information$_A$ results from a human agent's interaction with nature [17]. In some other definitions, information$_A$ may result from the cognitive activity of some non-human organism [18–20] or even some artificially intelligent agent [21,22]. However, in other cases, information$_A$ is claimed to exist independently in some world of ideas [23], or in other cases, information$_A$ may be seen through semiotic concepts [24]. Yet others claim that this information is data, or at least derived from data; from this perspective, the boundary between data and information$_A$ is quite fluid [4,5,25–30]. To many authors, information$_A$ is the prevalent conceptualization of information, and when they write about information, they claim that this is what information is really like [5,14,22,31–39] (note that this list is by necessity selective, and therefore incomplete).

---

[1] Mark Burgin [14,16] lists 32 mathematical formulas for information.
[2] We acknowledge that a generalization like this will never be 100% accurate, therefore some concepts of information may not fit into either of these two categories, or they may, in some sense, belong to both.
[3] By "abstract," we refer to something not existing in space-time as a physical object.

In the second category, information is regarded as a physical phenomenon, so we denote it as $information_C$, where the "C" represents "concrete".[4] As $information_C$ is a physical object or phenomenon, it can be measured and quantified, transformed, observed, and used. In this sense, such information is a property of matter and the physical world, and some claim it is the third constituent element of reality (e.g., [9]). $Information_C$ relates to the organization of nature, from the sub-atomic level to large-scale objects, from particles to cosmic bodies [40]. Authors who recognize the existence of physical information and its function in nature include the likes of von Weizsäcker [41], Turek [8], Hidalgo [10], Rovelli [11], Nagel [17], Devlin [34], Heller [42,43], Collier [44], Stonier [45], Polkinghorne [46], Seife [47], Schroeder [48,49], Dodig Crnkovic [50], Wilczek [51], Carroll [52], Davies [53], and Sole and Elena [54]. (As before, the list is by necessity selective and incomplete.) Table 1 below summarizes the main properties of $information_A$ and $information_C$.

Table 1. Main properties of $information_A$ and $information_C$.

	$Information_C$	$Information_A$
1	$Information_C$ is a physical phenomenon, so it exists objectively and is not relative to anything.	$Information_A$ is a cognitive agent's (artificial or biological) interpretation of physical stimuli, which may be a signal, the state of a physical system, or some other physical phenomenon.
2	$Information_C$ has no intrinsic meaning.	$Information_A$ exists for a cognitive agent, or it is relative to a cognitive agent. In other words, $information_A$ is agent-relative or ontologically subjective.
3	$Information_C$ is, in a sense, responsible for the organization of the physical world.	$Information_A$ has meaning for a cognitive agent. The cognitive agent may be a human, a biological system, or some artificially intelligent system.
4	For $information_C$, existence implies existence in the physical world, somewhere in the space–time continuum.	For $information_A$, existence denotes the presence of an abstract notion somewhere outside of space and time.

This study focuses on the second category of information, namely, $information_C$, and discusses how the existence[5] of this information is recognized, the properties that can be attributed to it, and its relationship to $information_A$ (see also Krzanowski [55]).[6] But why is the concept of $information_C$ so important, at least to the author? With the multiple definitions and classifications for information, our understanding of it resembles the parable of the blind men studying an elephant (Shah [56]): We each know a part of the investigated object, but not the whole. The hope is that the concept of $information_C$ may unify our view of information or put it in some order. Unifying the different concepts will further our understanding of reality and increase our recognition of its complexity and coherence, and such things are worth pursuing in their own right. That is why we search for a Grand Unified Theory (GUT) or a Theory of Everything (ToE).

Is $information_C$ a completely novel idea, though? Well, yes and no. On the one hand, the concept of information as a physical phenomenon is certainly not new. Information as a physical entity in various forms has been proposed by many authors (as mentioned above). On the other hand, no comprehensive study has investigated what the concept of $information_C$ entails or how it should be interpreted. This study, therefore, attempts to fill this gap.

In the following discussion, we first present how information as a physical phenomenon may be recognized and comprehended. The cited authors have supplied a short list of features that ground

---

[4] By "concrete", we mean existing in space and time as a physical object.

[5] In all cases of information, though, how can we say that both kinds of information exist? After all, the statement about how $information_A$ "exists" is not compatible with the statement about how $information_C$ "exists". The concept of existence in the case of information is discussed later in the study.

[6] $Information_A$ and $information_C$ have many interpretations that vary depending on what a specific author regards as concrete or abstract, so this division into two basic classes of information is certainly a generalization. There are also proposals for combining abstract and physical aspects into some sort of unified form of information. One example of this is Rovelli's purely physical meaningful information (i.e., physical information with meaning). We could denote such information as $information_{AC}$. The concept of $information_{AC}$ is not discussed here any further, however, but more details can be found in Rovelli's paper (Rovelli [11], Krzanowski [55]). Something worth noting, however, is that a concrete-abstract combination is only plausible after significantly changing the meaning of its component terms. In Rovelli's case, the concept of meaning (i.e., abstract knowledge) is reworked. In addition, the suggested resolution by Rovelli for the concrete–abstract concept of information has little to do with the general abstract–concrete problem of metaphysics [57,58]; Rovelli's proposal specifically addresses the concept of information, but it does not resolve the metaphysical abstract–concrete division.

the claims of this phenomenon's existence in physical facts. We then focus on three characteristics of information as a physical entity, although these characteristics are not necessarily novel, however, because every physical phenomenon must, by definition, possess some of these properties (see also [57–59]).[7] The question then is, therefore, one of whether we can attribute these features to information, not one of whether these features have been correctly chosen. Next, we look at the dependency between information$_C$ and information$_A$. We then propose that information$_A$ can be derived from information$_C$, but information$_A$ cannot be simply reduced to information$_C$. In the final section, we summarize the main points of this paper and present some problems that have been indicated for information$_C$, but left unresolved in the reviewed studies.

When we talk about concrete information (information$_C$) in this paper, we may use the term interchangeably with other terms like objective information (i.e., information that exists objectively) or physical information (i.e., information as a part of physical reality), or we may follow Rovelli's example of purely physical information (i.e., equivalent to physical information). Likewise, when we talk about abstract information (information$_A$), we may use the term interchangeably with similar terms like subjective information (i.e., information that depends upon a cognitive agent), epistemic information (i.e., information related to knowledge), and meaningful information (i.e., information conveying meaning). Finally, the generic term "information" is used to denote any specific concept that must be qualified according to a context or a descriptive term, otherwise it is devoid of meaning. In addition, when we talk about an object, we are referring to something that exists in physical reality.

2. Is Information$_C$ a Physical Phenomenon?

This section seeks to answer the following questions: Why do we claim that information is a physical phenomenon? Are there natural phenomena that we can class as information?

The concept of information is always associated with form, structure, or organization in some way. For example, information$_A$ represents meaning that a cognitive agent has associated with the form, structure, or organization of some entity. Information$_C$, in contrast, recognizes the form, structure, or organization of some entity as a purely physical phenomenon in itself. But what are the consequences of saying that information$_C$ is a physical phenomenon? It essentially means that information$_C$ is an irreducible aspect of physical reality. Information$_C$ is therefore not like a rainbow or a temperature of a volume of gas[8]. Instead, information$_C$ exhibits properties that we attribute to physical entities, namely, that (a) it is observable, (b) it is ontologically objective, (c) it can be manipulated, (d) it has no meaning, and I it can be quantified or measured.[9] Information$_C$ exists wherever physical reality exists, just like other physical phenomena. Information as a physical phenomenon that exists objectively has been recognized in the studies of, for example, Mynarski [9], Rovelli [11], Devlin [34], Heller [42,43], Collier [44], Stonier [45], Polkinghorne [46], Seife [47], Caroll [52], Dodig Crnkovic [50,60], and Dodig Crnkovic, Muller and Burgin [61]. Furthermore, Hidalgo [10], Stonier [45], and Seife [47], among others, posited that this information can be manipulated. VonWeizsäcker [41], Seife [47], Carroll [52], Stonier [45], Hidalgo [10], and Collier [44], meanwhile, recognized that this information is not associated with any meaning, and this observation has been confirmed in many other studies. In addition, the studies of von Weizsäcker [41], Barrow [62], Davies [53], Sole and Elena [54], and Heller [42,43] demonstrated that this information can be quantified. Thus, information$_C$ is not conceptualized as an abstract concept in the way that mathematical objects, ideas, or thoughts are abstract (i.e., they exist outside of space and time). Therefore, information$_C$ also does not belong to the

---

[7] The general features that characterize a physical phenomenon are often disputed, so we follow the most prevalent views on scientific realism [59].

[8] A rainbow and a temperature of a volume of gas are physically reducible to (can be explained by) more fundamental phenomena: white light refraction and the average kinetic energy of gas molecules, respectively.

[9] For more about the conditions for attributing ontology to a physical phenomenon, see the works of Worrall [59], Klee [63], Chakravartty [64], and Liston [65]. We bypass the discussion between the scientific realists and the anti-realists and assume the first position.

Platonic realm of Forms in either the old- or neo-Platonic sense (there will be more about information$_C$ and Plato's Forms later in the study). Information$_C$ is a natural phenomenon, and therefore, contained within nature, just as every physical object is (see also [63–65]).

To avoid making any unfounded claims, we state that information$_C$ must always exist within a physical medium (i.e., it cannot exist on its own). In other words, it is not something existing outside of physical objects, like Forms in Plato's Theory of Forms, but rather a part of physical reality.[10] From this perspective, information$_C$ is the third constituent element of nature [9,34,47], with the first two being energy and matter. This perspective, therefore, excludes any conceptualization of information$_C$ as something mysterious or esoteric or any other form of magical phenomenon hitching a ride on a physical medium.[11] What is more, we also prevent anyone from associating this information with some structure that overlays physical objects, as was suggested by Bates [66], Dinneen and Brauner [67], and in some sense Von Weizsäcker [41].

We now ask another question: What properties should we assign to information$_C$, and how should we interpret these properties?

3. Three Features of Information$_C$

When information is defined as a physical phenomenon, it has to have properties in common with other physical entities, namely, the ones that make them physical phenomena (e.g., Worrall [59], Klee [63], Chakravartty [64], Liston [65]). In addition, information$_C$ has a property that makes it a phenomenon in its own right. These properties are discussed below.

We posit that information$_C$ (1) exists objectively, (2) lacks meaning, and (3) expresses the organization or form of physical objects (we will further explain this last claim later on). We posit that we need all three of these features to describe information$_C$. First, objective existence alone is insufficient to uniquely characterize physical information, because everything in the real world exists objectively. The same applies to the lack of meaning and the presence of some organization when considered separately.[12] Any combination of two of the three features (e.g., lack of meaning with objectivity, organization with objectivity, etc.) is also not specific enough. Only the combination of these three features is sufficient to make information$_C$ a distinguishable physical phenomenon, so all three features are needed to adequately describe information$_C$. We do not claim that information$_C$ has only these features. But we only claim, based on our current understanding of information$_C$ these three features are sufficient to characterize it. It is possible that, in the future, this set of properties may change.

3.1. Information$_C$ Exists Objectively

If information is a physical phenomenon, it exists objectively. In other words, it has ontological objectivity [68]. Objectivity, in this context, refers to how the phenomenon in question exists independently of an observing agent, much like how rocks exist on the dark side of the Moon even though we cannot see them from Earth, or how the Earth existed for eons before sentient beings appeared.

Claiming that information$_C$ is a physical phenomenon is both necessary and obvious if we are to perceive information as part of the physical world. If this is the case, as we presume it is,

---

[10] Several authors explain the place of physical information in nature and propose a sort of matter–energy- information complex. As a conjecture, this proposal can remain, but as a scientific or even philosophical claim, it lacks enough specificity.

[11] The analogy of thermodynamic entropy may help here: Entropy itself is an abstract concept, but it refers to a real, physical phenomenon that can be measured, observed, and so on. The same is true of information$_C$: The concept is abstract, but it denotes a real physical phenomenon with specific properties.

[12] The common meaning of chaos should not be confused, as is often the case, with the meaning of chaos in nature. Chaos in nature (or what we call chaotic phenomena) is actually highly organized dynamic phenomena; entropy (incorrectly) that is often associated with chaos. It is incorrect to claim that these two phenomena are unstructured or disorganized. Nature is never disorganized—it is just the way it is. It only appears disorganized to us.

this information must be subject to the laws that govern the physical realm, although we are not exactly sure which laws apply to information$_C$ (e.g., the law of gravity, the law of conservation of energy, the three laws of thermodynamics, the theory of general relativity, or other yet unknown laws for information)[13].

The claim that information has a physical nature has been voiced in several studies. Von Weizsäcker claims that information has an "objective character," adding that information is related to the combination and composition of elementary particles. Heller, meanwhile, claims that information may be regarded as a material of the world, with the word "material" being used here in the sense of some physical substance from which things can be made, although Heller does not clarify synonyms like substance, stuff, or medium. Dodig Crnkovic, meanwhile, claims that information is an ontologically fundamental entity of physical reality, while Hidalgo proposes that what remains after separating meaning from meaningful information is just a physical phenomenon. Devlin claims that information is a part of reality that is on par with energy and matter, and Mynarski makes a similar claim. In a similar vein, Stonier claims that information has a physical reality, so it is an intrinsic feature of the universe. Finally, Seife claims that the world is composed of matter, energy, and information. In many other studies [9], the complex of matter, energy, and information is considered to be foundational in nature.[14]

3.2. Information$_C$ Has No Meaning

This claim states that information$_C$, as a natural/physical phenomenon, is meaning-less,[15] so the meaning is overlaid on, or added to, information$_C$ rather than it being an intrinsic part of it. How this meaning is created from, or associated with, information$_C$ is an open issue, however. Information$_C$ is just present out there, much like the rocks and the trees, regardless of whether or not some cognitive agent is there to observe them (see also [69–72].)[16] When a tree falls in the forest, it creates a wave of air pressure (i.e., an organized physical phenomenon), which in itself is meaningless. If we are close enough to hear, we add meaning to the sound. (Such a claim is easily verified through a first-hand experience)

Meaning is defined as something that has value or import for a cognitive agent, but this is just one possible definition. An agent may obtain something of value from information$_C$, something that has some significance for the agent's existence. The meaning may also be an interpretation of information$_C$—which otherwise, as we have said, is just an inert physical phenomenon—with us attributing to its various properties, such as structure, shape, or importance. Reading tea leaves and astrology are just two examples of such interpretations. The same information$_C$ may have different meanings for different agents, however, or it may not have any meaning at all for some agents. An agent is primarily an artificial or natural system that creates meaning for itself.

In principle, when perceiving environmental stimuli, some claim that there is no clear distinction between an artificial cognitive system, a simple biological system (e.g., a cell or a simple organism), and a conscious cognitive agent in terms of creating meaning [73–77]. From this perspective, and with a sufficient degree of abstraction, an agent could, in principle, be any organic or artificial system that

---

13 Every element of the physical realm is a subject to some physical laws, otherwise, by definition, it would not be physical.

14 The term "complex" denotes a combination of elements as "a whole made up of complicated or interrelated parts". We use this term to avoid referring to hylemorphism.

15 Meaning-less denotes an entity that lacks meaning in its essence, while meaningless denotes an entity that has no meaning in a specific context.

16 Meaning has many interpretations. For this study, if not otherwise stated, we follow the definition from the philosophy of language, where the term "meaning" denotes how language (linguistic constructs) relates to the world. A review of the various theories of meaning is beyond the scope and purpose of this work, however. An extensive list of references can be found in [69], among others. The theories claiming that meaning is the *correlata* to the world are contested by some good arguments by Chomsky [70–72].

senses and reacts to information$_C$ in some way.[17] The lack of any sharp boundaries between meanings in different agents creates a problem for identifying the locus of information$_A$, which is discussed later in this study.

3.3. Is Information$_C$ the Organization of Matter, or Is Responsible For It?

This claim asserts that information$_C$ is responsible for the organization of the physical world.[18] On the one hand, this seems to be the most critical aspect of information, because information, in general, is fundamentally associated with the concept of some form, organization, or structure. On the other hand, it is very difficult to define what it precisely means when we say that "information is associated with the concept of some form".[19] The difficulty arises from ambiguity in the concept of form and organization, as well as in the concept of "association".

Organization may be, as is commonly the case, interpreted as structure, order, form, shape, or rationality when it is perceived by a cognitive entity. We do not posit that information$_C$ is a structure in itself, because we do not know exactly what structure is, nor do we know what kind of structure would be associated with information or how this association would take place. Information$_C$ is certainly not the visible structure or shape of objects, although we admit that information$_C$ does reveal itself through the shape or structure of objects. Objects may be seen in a certain way, as being something, yet this may not be their essence, i.e., information$_C$ is responsible for the shape of things, but it is not their visible shape that is information$_C$. By way of an analogy, energy presents itself as the capacity of a system to do work, yet it is not work in itself; see the discussions in [78–83].

The word "association" denotes the relationship between the organization or form of information$_C$ and physical reality. This concept is critical to information$_C$, but we simply do not know what this relationship is. We only guess at its importance. More specific solutions always come down to, in some way, the matter–energy–information complex, but this relationship is as enigmatic as hylemorphism, both in its old and new renditions. The cited authors, despite referring to it, are also not clear about the meaning of this association. It seems that, for now at least, we need to remain at this rather imprecise descriptive level.

Why do we not associate information$_C$ with a specific structure, though? If we did try to equate information$_C$ with some domain-specific structure, such as a mathematical structure or the structure of the physical laws, information$_C$ would simply acquire the characteristics of that domain. In other words, such information would take on a mathematical or scientific flavor, respectively. It also seems that quantifications of information tend to quantify "sensible shapes" rather than the information itself. As an analogy, when we measure mass, we usually measure the gravitational force between it and the Earth rather than the actual mass.[20] In addition, stating that information$_C$ is a mathematical structure of reality would entail adopting some form of Neo-Platonism, which is not a generally

---

[17] Natural agents (i.e., biological systems) have been shaped by nature to sense its properties, including its structure (i.e., information$_C$). Nagel [17] discusses the dependency between an environment and an agent in detail. Indeed, we are built to interpret nature, so we could say that interpretation comes to us naturally (we are interpreters per se). Indeed, we seek interpretations, because they are essential for survival and because evolution deems that agents who fail to adequately perceive their environment will not survive.

[18] We are not sure how to interpret the function of information$_C$ in nature. One thing that is certain, however, is that the existence of information$_C$ is recognized in the form of objects. However, whether its role is that of Plato's Forms or Aristotelian *eidos*, or whether the role of information$_C$ is causal or not, is not well understood at present. Some studies claim [9] that information is a primary element of nature, or that information is a third element of nature in an energy–matter–information complex, but these are just intuitions. Due to this ambiguity about the role of information, the statements about information$_C$ and the organization of matter are imprecise.

[19] The structure/organization of physical reality is such a fundamental concept that it cannot be described through other concepts, because structure lies at the foundation of everything that exists. We cannot talk about reality without talking about structure. However, this is, of course, just conjecture.

[20] We stay away from unresolved disputes about the nature and ontology of mathematical constructs, because bringing unresolved disputes into the discussion will not further the resolution of other unresolved disputes, such as the nature of information$_C$ in this case.

accepted interpretation of mathematical objects (see e.g., [84]. We should, therefore, avoid these domain-specific claims and accept that information$_C$ is domain-neutral. What is more, if we were to just claim that information$_C$ is structure, it would not mean much, due to the ambiguous concept behind this unspecified structure. We would then ultimately end up treading the path of structural realism or informational structural realism with its epistemic or ontological versions. In addition, the structures in structural realism are passive, and they do not carry the meaning of "informare" (to shape), or at least nobody has attributed such causal powers to structures in the structural realism literature.

We must add here that different formal representations for information$_C$ are not incorrect (e.g., Shannon, Fisher, Chaitin), but they simply do not address the essence of what they measure[21]. This is how mathematical formula typically relate to nature [84].

Thus, we merely posit that information$_C$ is just one factor responsible for the organization of the physical world.

3.4. Information$_A$ and/or Information$_C$

We have differentiated two classes of information: abstract and concrete. In this section, we discuss how these two concepts relate to each other.

Information cannot be both abstract and concrete at the same time and in the same sense. What is abstract cannot be concrete, and what is concrete cannot be abstract, not when these terms are interpreted through their most common meanings.[22] Information$_C$ is a purely natural phenomenon—it is in nature, and it is a part of the physical world. Existence for this information implies the existence of a physical entity. Information$_A$, in contrast, is abstract, so it relies upon the existence of some cognitive system. (We exclude here Popper's concept of World 3 and similar ideas.) Now, let us explore these differences through an example.

Different physical structures may represent the same piece of music. It may be a series of air pressure waves with physical structure S_A, or it may be the grooves on a vinyl record with physical structure S_B. These and many other structures, can, under certain conditions, all be interpreted as the same piece of music (i.e., the same information$_A$). Thus, we have several physical objects, each with its unique organization, or information$_C$ (i.e., S_A, S_B, etc.), yet there is only one information$_A$, namely, the piece of music.[23] This piece of music is, therefore, clearly a common element of the objects with structures S_A, S_B, and so on. Obviously, this common element cannot be physical, however, because there is nothing physically common among these very different physical structures. Indeed, the only thing these physical objects share is how they all can be interpreted in the same way by someone or something (i.e., as the piece of music). As the music is not embedded in the physical structures, it must therefore exist outside (transcend) those physical structures in some sense. Thus, it would appear that information$_A$ clearly "exists" in some way. After all, how else could completely different physical structures convey the same meaning? It would seem clear to conclude that abstract information exists, so information is really information$_A$, not information$_C$. Such a conclusion has been reached in many studies [66,67,85,86]. However, we claim that this conclusion is incorrect or at least inaccurate. As controversial as it may seem to some, it seems that the proposed explanation for the piece of music is equally applicable to books, symbols, computer programs, game rules, works of art, and so on.

---

[21]   The case of the different mathematical representations of physical information is to some extend but not exactly, similar to the case of two different mathematical models of quantum mechanics (Schrodinger and Heisenberg). Herman Weyl stated that these models are "alternative representations of the same mathematical structure" (as quoted by Heller [86]). We do not claim that physical information is a mathematical structure but we suggest that the different mathematical models of information represent the same physical information as organization, as Schrodinger and Heisenberg's models are different representation of the QM structure.

[22]   Abstract things are objects outside space–time. Concrete objects exist in the physical world and are subject to the laws of physics [57,58].

[23]   These physical structures can also generally be converted from one to another (e.g., recording a radio performance onto a cassette tape) while preserving the capacity to be interpreted as the same piece of music (i.e., the same information$_A$).

The correct explanation, we believe, goes as follows: The physical structures (S_A, S_B, etc.) seem to carry the same information$_A$ because we, as cognitive agents, shape these structures in a certain way and later read/interpret them in a corresponding way, thus attributing an interpretation to them. These music-carrying physical structures are radically different, as are their physical carriers (e.g., airwaves, the impression of a vinyl record, etc.). What makes these carriers seem like they "carry" the same music is the interpretation of the agent rather than some factor (information$_A$) that exists outside (or transcends) the physical realm. The agent merely imposes (encodes) appropriate structures over a physical entity following some agreed-upon standard and later decodes it through an appropriate physical process, again following the agreed-upon standard. Music exists purely in the mind of the listener, composer, or interpreter, however, and "abstract information" does not float around in some metaphysical space, a sort of world of ideas that stands ready for us to access it.[24] As we said, information$_C$ is a multilevel organization of physical entities. In the case of music, we are interpreting the macro-level structures of physical objects, not their micro-level organization.

We need to be careful when using terms like "is in," "carries," "is embedded," "locked in," and "contains" when describing how information$_A$ relates to information$_C$. For example, information$_A$ is not "embedded" in a physical object in the usual sense of the word. Information$_A$ does not exist as some component, substratum, or ingredient of a physical object. Information$_A$ is created by a cognitive agent when it encounters some otherwise meaning-less physical phenomenon and decodes its form or organization (usually only a very selective subset of information$_C$) for its own use or benefit. This explanation should, we hope, counter any assertions that a physical object cannot be information because it cannot carry information$_A$ [67,85].

It is entirely possible that a different mind from some other world would interpret the same physical structure differently, so these structures actually only carry music for us. There is no music as we know it in viruses [87], biological cells [88], or heavenly bodies [89], yet these physical objects all have specific structures or organizations. We more or less understand the neural perceptual processes for perceiving the organization of physical objects through our cognitive systems, but we do not know how we perceive music. In short, we do not understand the phenomenology of music, nor do we understand the phenomenology of information$_A$ in general.

We assume here that the mind is a non-reducible biological phenomenon rather than a reducible emergent phenomenon, even if we cannot accurately explain what it is at our current level of knowledge (apparently, some animals also enjoy Mozart or Bach [90]). We obviously try to avoid Descartes' duality, so we do not postulate the existence of abstract information as some esoteric entity. We also do not advocate the strict reductionism of mental features to neuronal levels. Of course, this argument can apply only if the concept of the mind is positioned as the locus of information$_A$, which is created (in part) based on different physical signals/stimuli [91]).[25] While this explanation takes away the "abstractness" of information$_A$ (because it grounds such information in a biological system), it does not explain how the information in our minds is created, transformed, stored, accumulated, and "transferred" back into our artifacts when we shape physical things. For example, there is no natural law that shapes a car, a table, or a watch (see the modern version of Paley's argument in [92]).[26] We must admit that we simply do not yet understand how the mind works, so our theories of the mind cannot provide a conclusive explanation for it at present.[27] The only thing we can assume is that the natural world, including the mind and its created artifacts, is causally closed, and there is no "bifurcation" of nature into the world of nature and the world beyond, as implied by Descartes but denied by Whitehead

---

[24] Apparently, Mozart claimed that he did not compose the music but merely noted it down, however this may be a psychological phenomenon rather than a scientific argument for the independent existence of information$_A$.

[25] See, for example, "human behavior is determined by physical processes in the brain" [91]. Similar views are widespread in the literature of the field.

[26] I am referring to the 747 junkyard argument, the details of which may be found in [92].

[27] We assume that the current explanations of the mind as a kind of software and the brain as a kind of hardware are wrong and misguided. See, for example, the arguments of Searle [94].

and other modern philosophers of science. What is more, we may safely claim that thoughts and the mind are part of this world, much like the other phenomena we experience and observe. This claim, however, is not meant to endorse material monism and certainly not physical reductionism [93,94]. There are philosophies of the mind (e.g., biological naturalism) that classify the mind, consciousness, thoughts, ideas, and so on as not being reducible to matter, yet they have a material base [68,95–99] (note that many philosophers of the mind contest this antireductionist view, such as [100]).

In summary, we claim that information$_C$ exists objectively in the physical realm. It is unique in the sense that every physical object has its own organization or structure. Two physical objects may have a similar organization, or their organization may mean the same thing to us (as in the example of music), but their sameness (to us) does not come from these objects. Information$_A$ also exists (as an abstract entity), but its existence is contingent upon the presence of a mind or other cognitive system. Information$_A$ can also be transferred, stored, communicated, preserved, and transformed, but it needs a physical media to carry it and a mind to recreate it.

4. Information$_C$: To Be or Not to Be?

Thus far, we have presented research that recognizes the existence of information$_C$, but some authors deny that such information exists. We look at some of these studies and weigh their arguments.

Dinneen and Brauner [67] claim that "information-as-a-thing," which for them is information as a physical phenomenon, cannot account for "typical views of information".[28] These problems are avoided, they say, if information is seen as an abstract entity. The only example of "information-as-a-thing" (i.e., physical information) they provide in their 2018 study is a book (as a physical object). We could delve deeper into Dinneen and Brauner's argument, but this is not necessary. It seems that they set up their definition of physical information to fail, because according to their definition, physical information (a book) cannot be information because it cannot have meaning, so information is not physical. This is rather obvious, though. Physical information, or a book in their example, is meaningless by definition (a book is a physical object). Dinneen and Brauner's claim is in some way correct, because the physical object (the book) is meaningless in itself, as all physical objects are. However, their argument against the existence of physical information as a physical phenomenon, based on the example of the book, is incorrect because they are looking for meaning where there is none to be found. Dinneen and Brauner's attempt was, therefore, certain to fail, because they were looking for meaningful information or meaningful physical information rather than just physical information. For Dinneen and Brauner, "typical views of information" reflects what we refer to in this study as information$_A$. There is nothing typical about this, even though it may be the most prevalent view of information. In science, however, the truth of a theory is not determined according to a majority vote, and the minority opinion is often the correct one. Dinneen and Brauner, it seems, missed the nature of information$_C$; information$_C$ is not a physical object itself, but its organization (in a sense) discussed here.

In their earlier paper, Dinneen and Brauner [85] formulate three arguments for why a physical thing cannot be information or, more precisely, why what they call "information-as-a-thing" cannot exist. First, "the value of the physical representation is first and foremost its content, and not the physical embodiment of it". Thus, putting forward information-as-a-thing as information is clearly ignoring the content of a physical thing, and we are concerned with this content. Second, talking about physical objects as information is not accurate, because when talking about information, we are more interested in what these physical objects (e.g., DVDs, CDs, USB sticks) contain rather than the things themselves. Thus, the definition of information as a physical object is misleading, as well as the conclusion of the first argument. Third, the same physical object may contain different information depending on time and place: For example, a book's content may be interpreted differently. This creates, according to Dinneen and Brauner, a metaphysical problem of identity. If information is a physical

---

28 " ... that physical things cannot be information, and information therefore cannot be a physical thing" [67].

thing, it must be the same in all circumstances, otherwise we would have two or more things being the same physical object. We have partially addressed these three arguments in Section 4. As in the above discussion, Dinneen and Brauner do not distinguish between information$_C$ and information$_A$. When they talk about information, they are actually talking about information$_A$. They are, in a sense, attributing information$_A$ with a physical presence, and as we have pointed out many times in this study, information$_A$ is not physical, and a physical object is not information$_A$. It may "contain" information$_A$ for one or more agents, at least in the sense of "contain' as explained above, but a physical object is never information$_A$. In a rather stretched analogy, we could say that energy does not work, nor does it contain work, but it is certainly related to work. The analogy stops here, though. In defining information (information$_A$) as a physical object or information-as-a-thing, we are obviously making a mistake by conflating the abstract with the concrete, which will clearly never work. As we said before, physical information is not a physical object in the sense of specific object like a book, a DVD, etc. Physical information, or information$_C$, is the organization of these objects, as explained in the previous sections. What Dinneen and Brauner face is the concrete-abstract split indicated by Davies [53] and Rovelli [11],[29] but while Rovelli and Davies managed to comprehend and overcome it, Dinneen and Brauner did not.

Bates [66], following Edwin Parker (quoted by Bates), identifies information in nature as a pattern of, or within, physical things. However, Bates' information is not physical, because it is a pattern, an abstract concept realized through a physical medium and recognized by a cognitive agent. This interpretation is seen in Bates' claims that while information as a pattern is everywhere in the universe, total entropy is pattern-free,[30] so it has no information. Therefore, according to Bates, total entropy cannot be interpreted as a pattern, so Bates' notion of "information as a pattern of physical things" is added to some physical phenomena, but not to others.[31] In short, we may say that in Bates' view, information is not a physical entity, even when it is associated with physical objects, but rather a perceived pattern of physical objects. In her own example, some physical phenomena are information-free. It therefore seems that Bates' information has nothing to do with information$_C$ and is more akin to the concept of natural information seen in Millikan's work [101], which is also not information$_C$. By Millikan's very definition, natural information comprises infosigns carried by natural phenomena that "initiate perception" [101]. In this definition, natural information appears to be simply information$_C$ plus the meaning or interpretation for a physical carrier. Recall that information$_C$ does not need to initiate perception to exist. A similar definition of natural information is given by Piccinini and Scarantino [102].

The conviction that information must have meaning has prevented many researchers from recognizing the existence of physical information. This "epistemic turn" (see James [103]).[32] is characteristic of modern philosophy, and it began with Descartes. For example, von Weizsäcker [41] and others later on, claimed that information must be also physical in some way, yet he could not recognize information without meaning.

Some of the arguments against the concept of information$_C$ have been generated by identifying information$_C$ with Plato's Forms. One such argument, namely, a modified version of the Third Man argument, asks: If information$_C$ is in every physical object, is information$_C$ in information$_C$? Another argument questions how the same information$_C$ may exist in different physical objects at the same

---

[29] The problem is stated as follows: "How can information be physical and abstract at the same time?"

[30] We do not go into details about what is "total entropy" or whether information as a pattern would appear if entropy was less than total (whatever that means for Bates) (i.e., would information as a pattern disappear at one point, or would it appear or disappear gradually?).

[31] The claim that "total entropy is pattern-free" is incorrect, because every physical phenomenon has some organization or pattern, although it may be beyond our understanding in some cases. Bates repeats the common misconception of equating entropy (assumedly thermodynamic entropy) with the popular notion of chaos (of sorts).

[32] The "epistemic turn" denotes the reorientation of modern philosophy from ontology to epistemology as the main philosophical perspective on nature.

time (i.e., how does the same physical thing (information$_C$) exist in many different places at the same time?) These problems apply to Plato's Forms in his metaphysical view, but as we said from the start, information$_C$ is not one of Plato's Forms, because such objects exist outside space and time, so in this sense, they are abstract objects. While Plato's Forms are in some way physical things (in Plato's view), the nature of their existence (outside the space-time) and relation to reality is exactly what makes them controversial. Information$_C$ as a physical phenomenon does not suffer from these shortcomings, just as physical objects do not suffer from the same shortcomings. For example, we do not question whether energy is within energy or whether matter is within matter, even though these phenomena are everywhere. Information$_C$ is more akin to the Aristotelian concept of eidos, but as we pointed out earlier, such analogies to ancient ideas are very precarious and should be drawn with great restraint. This is why we do not discuss them further in this study, or we do not propose them to be renditions of information$_C$.

Taking a larger view, any researcher who claims that the concept of information is inherently and exclusively associated with meaning and knowledge is implicitly denying the existence of information as a physical phenomenon for the obvious reasons explained above (i.e., information cannot be both abstract and concrete in the same way and at the same time). Surprisingly, such claims are made with full knowledge that human agents are physical-information-processing systems (see, for example, [18,19,104–106]), and as a computer, our main data-processing system is a purely physical, mindless, and meaningless device. We need to take us (or the mind) out of this picture in a kind of Copernican move to see information$_C$.[33]

As a reminder, the existence of information as a physical entity is supported by the studies in which information (the concept of information) has been found to have properties that are attributable to physical objects, the studies that have found information useful for explaining certain physical processes, and the studies that have found information as a unifying factor in explaining a range of natural phenomena. (See the authors quoted in the earlier sections of this study.)

5. Physical Information Revisited: Conclusions and Questions

The time has come to summarize the main findings of this study. We conceptualized physical information, or information$_C$, as a natural phenomenon that has three properties: (i) a physical, objective existence, (ii) the absence of intrinsic meaning, and (iii) an organization of, or within, nature. These properties are, of course, subject to many interpretations, so they need to be understood within the context of the cited studies.

Information$_C$, (as a physical phenomenon) exists objectively in the same sense as the physical world around us. Information$_C$ is not abstract in the way that mathematical concepts and ideas are abstract. However, whatever exists contains information$_C$ in some form, and there is no physical phenomenon without information$_C$, because every physical phenomenon has some level of organization (even if we do not recognize it). Information$_C$ is meaning-less, however, just like all other physical phenomena. Meaning is associated with, or attributed to, information$_C$ by some sort of cognitive system. How meaning is created, though, lies beyond the scope of this study. Information$_C$ is a constituent element of nature, and it discloses itself through the organization of the physical world. 'Organization' is a fairly broad concept, but in this study, organization can be regarded as structure, order, form, or shape, although it cannot be simply identified with it.

Information$_C$ is a carrier (in the sense that was explained earlier) of information$_A$, which is information with meaning or value. Meaning is what we, or generally any cognitive agent, associate with information$_A$. The process of creating information$_A$ depends upon the cognitive agent, and in some sense, information$_A$ exists in the agent in the same way that thoughts and ideas exist.

---

[33] By "a Copernican move' we understand the position in which a human person is not the vantage point from which to look at nature.

We always need to be careful when using the term "exist," however, because it means different things for abstract and concrete information. Meaning is defined as something of value for the cognitive agent. From information$_C$, an agent may derive something that has some significance for the agent's existence or functions, so the agent essentially creates meaning for itself. The same information$_C$ may have a different meaning for a different agent, or more significantly, the same information$_C$ may have no meaning at all for an agent. What constitutes an agent, as well as which agents can create and possess abstract information, is disputable. We could represent the meaning of information on a linear spectrum from purely physical information at one extreme to fully meaningful information for a human agent at the other extreme. The precise boundary between concrete and abstract information may seem fluid, because the concept of meaning is often extended to minimally cognitive organisms, whatever that means, and artificial systems [107,108]. (Indeed, the fluidity of this boundary was exploited by Rovelli in his concept of information$_{AC}$.) However, one pole of this meaning spectrum is meaning-less, and this is where information$_C$ lies. The problems with information$_A$ vs. information$_C$ begin when we start trying to attribute meaning to information$_C$.

The fact that we do not propose any mathematical or formal formula for information$_C$ derives from the fact that this information is represented as an organization or structure of nature, so it may be quantified at different levels in many different ways through many different mathematical models, yet there is no single preferred way to do this. Some models may seem more useful than others, so they may acquire an aura of being "the right model".

Last but not least, we may ask how relevant Floridi's General Theory of Information (GTI) is to the concept of information$_C$ [4]. We ask this because the GTI is seen as an exemplary, comprehensive attempt to define what information is and subsume most, if not all, of the other definitions [85]. Very briefly, the GTI is defined as "data + meaning". In other words, GTI information is data endowed with meaning. We have already discussed what meaning is, but what is data here? Floridi is not clear, however. He writes: "nothing seems to be a datum per se. Rather, being a datum is an external property". This definition somewhat expresses the idea that a datum does not exist in its own right but rather that it is some "X" with added meaning (i.e., it is an external property). Thus, the concept of a datum in the GTI is relative, and its existence depends on some agent elevating it to the status of a datum. Thus, the GTI may be alternatively expressed as "(X + meaning) + meaning". This X is defined as a "fracture in the fabric of being," a "lack of uniformity," or an "external anchor of information". In fact, we do not know what it is, and we cannot by definition know what it is (Floridi compares it to Kant's noumena). So, the GTI is essentially grounded in something we do not know. A somewhat charitable interpretation of the GTI (by disregarding the status of X) would be that because GTI information is tightly coupled with meaning, it actually has nothing to do with information$_C$. The GTI information is, therefore, a comprehensive, very detailed formulation of information$_A$, but as we said, this would be a charitable interpretation. Notwithstanding the grounding problem (the status of X), the GDI is still a very useful and thorough attempt to organize and rationalize information$_A$ and may serve as a reference point in the discussions about the nature of information.

Maybe we should again emphasize what is proposed in this study. Every physical phenomenon has some organization or form (in the sense explained), and we denote this as information$_C$. Information$_C$ is a concrete, physical phenomenon, thus it has no meaning. Cognitive agents that interact with nature—or as we could instead say, the physical environment—sense and react to information$_C$. The interaction between an agent and the environment is physical. In some cases, an agent absorbs some subset of information$_C$ via its sensory apparatus, and on combining this with its internal resources in some way, it creates information$_A$. At present, the process in which information$_A$ is created is only vaguely understood. (We do not accept any views that would reduce the mind to neuronal levels or explain the mind as an illusion, a kind of software, or other emergent phenomenon in a reductive sense.) A physical object (a book, a DVD, a CD disk as in the examples cited) is not information$_C$, although information$_C$ is part of it, and in principle, it does not carry any information$_A$. Unless we were to postulate some transcendent ontology for meaningful entities (see Popper or Peirce), information$_A$ does not exist in a physical sense. Information$_A$ and information$_C$ may coexist, but existence in both cases means a different thing. The conflict arises when we try to conflate information$_A$ and information$_C$ or attribute the same mode of existence to them.

Several questions about the concept of information$_C$ remain unresolved, and the proposed list below presents some of them. These questions are speculative, but they appear in the research on information$_C$, so they are related to this work. Of course, the real list of unresolved questions about the nature of physical information is likely to be much longer than the one below.

In all the questions below, the term "information" refers to information$_C$.

Question 1: Do laws for the conservation of information exist, and if they do, what do they claim? Is the total amount of information in the universe therefore constant? *This question probes the problem of "the conservation of information". If information is fundamental to whatever exists in the physical world, does it follow laws for its preservation, much like energy?* (Suggested by the writings of Carroll [52], for example.)

Question 2: Can we claim that whatever exists must contain information$_C$? Can we defend the paninformatism claim that information is everything that exists? What is more, is paninformatism related to panpsychism? *This question probes the claim that information is in everything that exists. Can such a claim be justified? And does such a claim amount to some kind of paninformatism or panpsychism? If so, what precisely would this entail? Would such a claim trivialize the concept of information?* (Suggested by the writings of Stonier [45], Turek [8], and Carroll [52], for example.)

Question 3: Can we interpret information$_C$ as a causal factor, and how could such a claim be verified? *This question probes the alleged causal role of information in the physical world. It amounts to the question of whether information is a passive or active element in nature and what the nature of this activity would be.* (Suggested by the writings of Carroll [52], and von Weizsäcker [41], for example.)

Question 4: Information$_C$ is foundational to the physical universe, but in what sense can this statement be made? *This question probes the claim that information is fundamental to nature, but what exactly would this mean? Should such a claim be interpreted along the lines of the proposed information–matter–energy complex? Or should it be interpreted more metaphysically like the Logos of The Bible or the Tao of Tao-Te-Ching as an all-pervading and primordial element of existence?* (Suggested by the writings of Heller [42,43], Dodig Crnkovic [60], Stonier [45], for example.)

Question 5: Can we say that highly complex and chaotic (i.e., non-linear, dynamic) systems have no information$_C$? *This concerns the problems of chaos and non-linear, dynamic systems. Does information play a role in such systems? Quite often, chaos is associated with a lack of information, which seems to be a questionable interpretation of a physical phenomenon.* (This issue was indicated by Bates [2].)

Question 6: Does information$_C$ imply some form of modern hylemorphism?[34] *This question seeks to identify the similarities between information and hylemorphism in its modern interpretations. The problem of the nature of information and matter and energy has resurfaced in the works of many authors (see the references in this paper), and they all seem to echo Aristotelian metaphysics (see Jaworki [109])* (Suggested by the writings of Polkinghorne [46], Turek [8], Krzanowski [110], and Carroll [52], for example.)

Question 7: Does the fact that information is physical change the meaning of computation from one of symbolic processing to processing physical information? *We associate computation with*

---

[34] William Jaworski argues why the hylemorphic structure is the best, and perhaps only, means for explaining the persistence of individuals who change their matter over time. Hylemorphism claims that some individuals, paradigmatically living things, are composed of physical materials with a form or structure that is responsible for them existing and persisting as the kind of things they are. One objection to hylemorphism is that an account of the physical materials that comprise an individual is insufficient to account for everything it is and everything it does. William Jaworski, however, argues that this objection fails insofar as hylemorphic structure is the best, and perhaps only, means for explaining the persistence of individuals who change their matter over time [109]. A similar claim was made almost 40 years earlier by Turek in a 1978 article on the concept of information and its relation to a restricted form of hylemorphism [8].

symbolic processing, but computation in computers is, in fact, a highly structured, pure physical process (e.g., as Searle said, "computation is in the eye of the beholder"). Could we extend the concept of computation to any physical process involving changes in physical organization without trivializing the concept of computing? Do we even care? (Suggested by the writings of Seife [47], Dodig Crnkovic [50,60], and Dodig Crnkovic and Mueller [61], for example.)

Question 8: Can information be equated to some kind of structure, and what would this mean for the concept of structure? *This question proposes explaining the concept of information$_C$ through the concepts of structure and structural realism.* (Suggested by the writings of Heller [42,43], and Schroeder [48,49].)

Funding: This research received no external funding.

Conflicts of Interest: The author declares no conflict of interest

References

1. Wersig, G.; Neveling, U. The Phenomena of Interest to Information Sciences. Available online: https://sigir.org/files/museum/pub-13/18.pdf (accessed on 20 April 2019).
2. Bates, M.J. Fundamental forms of information. *J. Am. Soc. Inf. Sci. Technol.* **2006**, *57*, 1033–1045. [CrossRef]
3. Janich, P. *What is Information*; University of Minnesota Press: London, UK, 2006.
4. Floridi, L. Philosophical Conceptions of Information. In *Intelligent Tutoring Systems*; Springer Science and Business Media LLC: Berlin/Heidelberg, Germany, 2009; Volume 5363, pp. 13–53.
5. Floridi, L. *The Philosophy of Information*; Oxford University Press (OUP): Oxford, UK, 2010.
6. Nafria, J. What is Information? A Multidimensional Concern. *TripleC* **2010**, *8*, 77–108. [CrossRef]
7. Adriaans, P. Information. *The Stanford Encyclopaedia of Philosophy*. Edward, N., Zalta, D., Eds.; Available online: https://plato.stanford.edu/archives/spr2019/entries/information/ (accessed on 2 November 2019).
8. Turek, K. *Filozoficzne Aspekty Apojęcia Informacji*; Zagadnienia Filozoficzne w Nauce: Krakow, Poland, 1978; pp. 32–41.
9. Mynarski, S. *Elementy Teorii Systemow i Cybernetyki*; Panstwowe Wydawnictwo Naukowe: Warszawa, Poland, 1981.
10. Hidalgo, C. *Why Information Grows*; Penguin Books: London, UK, 2015.
11. Rovelli, C. Meaning = information + evolution. *arXiv* **2016**, arXiv:1611.02420v1. Available online: https://arxiv.org/pdf/1611.02420.pdf (accessed on 28 March 2020).
12. Shannon, C.E. A Mathematical Theory of Communication. *Bell Sys. Tech. J.* **1948**, 379–423. [CrossRef]
13. Chaitin, G. *Algorithmic Information Theory*; Cambridge University Press: Cambridge, UK, 2004.
14. Burgin, M. Information: Problems, Paradoxes, Solutions. TripleC.1.1. 2003, pp. 53–70. Available online: http://tripleC.uti.at/ (accessed on 2 November 2017).
15. Wittman, R. Fisher Matrix for Beginners. [Online]. 2018. Available online: Wittman.physics.ucdavis.edu/Fisher-matrix-guide.pdf (accessed on 1 December 2018).
16. Burgin, M. Is Information Data? 2005. Available online: http://www.mdpi.org/fis2005 (accessed on 2 November 2018).
17. Nagel, T. *Mind and Cosmos: Why the Materialist Neo-Darwinian Conception of Nature is Almost Certainly False*; Oxford University Press: Oxford, UK, 2012.
18. Maturana, H.R. Biology of Cognition. Reprinted in Maturana and Valera. *Construct. Foundat.* **2011**, *6*, 352–362.
19. Maturana, H.R.; Varela, F.J. *Autopoiesis and Cognition*; Springer Science and Business Media LLC: Berlin/Heidelberg, Germany, 1980; Volume 42.
20. Maynard Smith, J. The Concept of Information in Biology. *Philos. Sci.* **2000**, *67*, 177–194. [CrossRef]
21. Vernon, D.; Metta, G.; Sandini, G. A Survey of Artificial Cognitive Systems: Implications for the Autonomous Development of Mental Capabilities in Computational Agents. *IEEE Trans. Evol. Comput.* **2007**, *11*, 151–180. [CrossRef]
22. Vernon, D.G. *Artificial Cognitive Systems. A Primer*; The MIT Press: Cambridge, UK, 2014.

23. Popper, C. Three Worlds. The Tanner Lecture on Human Values. Delivered at The University of Michigan. 1978. Available online: https://tannerlectures.utah.edu/_documents/a-to-z/p/popper80.pdf (accessed on 7 April 2019).
24. Peirce, C.S. *Peirce on Signs: Writings on Semiotic*; James, H., Ed.; University of North Carolina Press: Chapel Hill, NC, USA, 1994; p. 294.
25. Ackoff, R.L. From data to wisdom. *J. Appl. Sys. Anal.* **1989**, *16*, 3–9.
26. Davenport, T.H. *Information Ecology*; Oxford University Press: Oxford, UK, 1997.
27. Drucker, P. *Management Challenges for the 21st Century*; Informa UK Limited: Colchester, UK, 2012.
28. Saint-Onge, H. Linking Knowledge to Strategy. In Proceedings of the Strategic Planning for KM Conference, Toronto, ON, Canada, 28–29 May 2002; pp. 28–29.
29. Terra, J.L.; Angeloni, T. Understanding the difference between Information Management and Knowledge Management. Available online: http://providersedge.com/docs/km_articles/Understanding_the_Difference_Between_IM_and_KM.pdf (accessed on 28 March 2020).
30. Drogan, J. Data, Information, And Knowledge—Relevance and Understanding. 2009. Available online: http://static1.1.sqspcdn.com/static/f/23196/12369003/1306255277903/Data+Information+and+Knowledge+-+Relevance+and+Understanding.pdf?token=qXSZz6nhIKiwutGvrXNIeIQjm6M%3D (accessed on 12 November 2019).
31. Bar-Hillel, Y.; Carnap, R. Semantic Information. *Br. J. Philos. Sci.* **1953**, *4*, 147–157. [CrossRef]
32. Brooks, B. The foundations of information science. Part I. Philosophical Aspects. *J. Inform. Sci.* **1980**, *2*, 125–133. [CrossRef]
33. Buckland, M.K. Information as thing. *J. Am. Soc. Inf. Sci.* **1991**, *42*, 351–360. [CrossRef]
34. Devlin, K. *Logic, and Information*; Cambridge University Press: Cambridge, UK, 1991.
35. Sveiby, K. E. What is Information. 1994. Available online: http://www.sveiby.com/articles/information.html (accessed on 20 April 2016).
36. Losee, R.M. A discipline independent definition of information. *J. Am. Soc. Inf. Sci.* **1997**, *48*, 254–269. [CrossRef]
37. Dretske, F. *Knowledge and the Flow of Information*; CSLI Publications: Cambridge, UK, 1999.
38. Casagrande, D. Information as Verb: Re-conceptualizing Information for Cognitive and Ecological Models. *J. Ecol. Anthr.* **1999**, *3*, 4–13. [CrossRef]
39. Lensky, W. Information: Conceptual Investigation. *Information*. 2010, pp. 74–118. Available online: www.mdpi.com/journal/information (accessed on 6 October 2015).
40. Dittrich, T. 'The concept of information in physics': An interdisciplinary topical lecture. *Eur. J. Phys.* **2014**, *36*, 15010. [CrossRef]
41. von Weizsäcker, C. *Die Einheit der Natur, Munchen: Verlag, Berlin*; PIW: Warszawa, Poland, 1971.
42. Heller, M. *Ewolucja Pojęcia Masy*; Heller, M., Michalik, A., Mączka, J., Eds.; Filozofować w kontekście nauki. PTT: Krakow, Poland, 1987; pp. 152–169.
43. Heller, M. *Elementy Mechaniki Kwantowej dla Filozofow*; Copernicus Center Press: Krakow, Poland, 2014.
44. Collier, J. Intrinsic Information. In *Information, Language and Cognition: Vancouver Studies in Cognitive Science*; Hanson, P.P., Ed.; University of British Columbia Press, now Oxford University Press: Oxford, UK, 1990; pp. 390–409.
45. Stonier, T. *Information and the Internal Structure of the Universe*; Springer Science and Business Media LLC: Berlin/Heidelberg, Germany, 1990.
46. Polkinghorne, J. *Faith, Science & Understanding*; Yale University Press: New Haven, UK, 2000.
47. Seife, C. *Decoding the Universe*; Viking: London, UK, 2006.
48. Schroeder, M. Philosophical Foundations for the Concept of Information: Selective and Structural Information. 2005. Available online: www.mdpi/fis2005 (accessed on 12 November 2019).
49. Schroeder, M.J. Ontological study of information: Identity and state. *Kybernetes* **2014**, *43*, 882–894. [CrossRef]
50. Dodig, C.G. Alan Turing's Legacy: Info-Computational Philosophy of Nature. [Online]. 2012. Available online: http://arxiv.org/ftp/arxiv/papers/1207/1207.1033.pdf (accessed on 7 October 2015).
51. Wilczek, F. *A Beautiful Question*; Penguin Books: London, UK, 2015.
52. Carroll, S. *The Big Picture on the Origins or Life, Meaning and the Universe Itself*, 1st ed.; OneWord: London, UK, 2016.
53. Davies, P. *The Demon in the Machine*; University of Chicago Press: Chicago, IL, USA, 2019.

54. Solé, R.V.; Elena, S.F. *Viruses as Complex Adaptive Systems*; Princeton University Press: Princeton, NJ, USA, 2019.
55. Krzanowski, R. Does purely physical information have meaning? A comment on Carlo Rovelli's paper: Meaning = information + evolution. *arXiv* **2004**, arXiv:1611.02420.
56. Shah, I. *Tales of the Dervishes*; Octagon Press: London, UK, 1993.
57. Rosen, G.F.; López, J.; Martínez-Vidal, M. *Abstract Objects, The Stanford Encyclopedia of Philosophy*. Edward, N., Ed.; Available online: https://plato.stanford.edu/entries/abstract-objects/ (accessed on 2 November 2019).
58. Armstrong, D.M. *Sketch for a Systematic Metaphysics*; Oxford University Press (OUP): Oxford, UK, 2010.
59. Worrall, J. Scientific Realism and Scientific Change. *Philos. Q.* **1982**, *32*, 201. [CrossRef]
60. Dodig, C.G. *The Info-Computational Nature of Morphological Computing*; *Philosophy and Theory of Artificial Intelligence*; Müller, V., Ed.; Studies in Applied Philosophy, Epistemology and Rational Ethics; Springer: Berlin/Heidelberg, Germany, 2013.
61. Dodig-Crnkovic, G.; Müller, V.C.; Burgin, M. A Dialogue Concerning Two World Systems: Info-Computational vs. Mechanistic. In *Information and Computation*; World Scientific: Singapore, 2011; pp. 149–184.
62. Barrow, J.D. *New Theories of Everything*; Oxford University Press: Oxford, UK, 2007.
63. Klee, K. An Introduction to the Philosophy of Science. *Philos. Books* **1997**, *31*, 59–61. [CrossRef]
64. Chakravartty, A. *"Scientific Realism", The Stanford Encyclopedia of Philosophy*. Edward, N., Ed.; 2017. Available online: https://plato.stanford.edu/entries/scientific-realism/ (accessed on 2 November 2017).
65. Liston, M. *Scientific Realism and antirealism*. 2019. Available online: https://www.iep.utm.edu/sci-real/ (accessed on 23 June 2019).
66. Bates, M.J. Information, and Knowledge: An Evolutionary Framework for Information Science. Available online: https://pdfs.semanticscholar.org/15cc/84f8486b1fd1f71a2a899015ab0e0cb86330.pdf?_ga=2.168931681.191364369.1592979337-273105443.1590587312 (accessed on 27 April 2020).
67. Dinneen, J.D.; Brauner, C. Information-not-thing: Further problems with and alternatives to the belief that information is physical. In Proceedings of the Annual Conference of CAIS/Actes du Congrès Annuel de l'ACSI, Toronto, ON, Canada, 31 May–2 June 2017; University of Alberta Libraries: Edmonton, AB, Canada, 2018.
68. Searle, J. *Intentionality: An Essay in the Philosophy of Mind*; Cambridge University Press: Cambridge, UK, 2008.
69. Speaks, M. *"Theories of Meaning", The Stanford Encyclopedia of Philosophy*. Edward, N., Ed.; Available online: https://plato.stanford.edu/entries/meaning/ (accessed on 20 April 2019).
70. Chomsky, N. On Mind and Language. Lecture online. 2013. Available online: https://www.youtube.com/watch?v=3DBDUIDA3t0 (accessed on 12 November 2019).
71. Chomsky, N. The Concept of Language. Lecture online. 2014. Available online: https://www.youtube.com/watch?v=hdUbIlwHRkY (accessed on 12 November 2019).
72. Chomsky, N. Language, Creativity, and the Limits of Understanding. Lecture online. 2016. Available online: https://www.youtube.com/watch?v=XNSxj0TVeJs (accessed on 12 November 2019).
73. Adleman, L.M. Computing with DNA. *Sci. Am.* **1998**, *279*, 54–61. [CrossRef]
74. Church, G.M.; Gao, Y.; Kosuri, S. Next-Generation Digital Information Storage in DNA. *Science* **2012**, *337*, 1628. [CrossRef] [PubMed]
75. Goldman, N.; Bertone, P.; Chen, S.; Dessimoz, C.; Leproust, E.M.; Sipos, B.; Birney, E. Towards practical, high-capacity, low-maintenance information storage in synthesized DNA. *Nature* **2013**, *494*, 77–80. [CrossRef] [PubMed]
76. Zhirnov, V.; Zadegan, R.; Sandhu, G.S.; Church, G.M.; Hughes, W.L. Nucleic acid memory. *Nat. Mater.* **2016**, *15*, 366–370. [CrossRef]
77. Erlich, Y.; Zielinski, D. DNA Fountain enables a robust and efficient storage architecture. *Science* **2017**, *355*, 950–954. [CrossRef] [PubMed]
78. Iona, M. Letters: Energy is the Ability to do Work. *Phys. Teach.* **1973**, *11*, 259. [CrossRef]
79. Lehrman, R.L. Energy Is Not the Ability to Do Work. *Phys. Teach.* **1973**, *11*, 15. [CrossRef]
80. Hicks, N. Energy is the capacity to do work-or is it? *Phys. Teach.* **1983**, *21*, 529. [CrossRef]
81. Hobson, A. Energy and Work. *Phys. Teach.* **2004**, *42*, 260. [CrossRef]
82. Coelho, R.L. On the concept of energy: History and philosophy for science teaching. *Proc. Soc. Behav. Sci.* **2009**, *1*, 2648–2652. [CrossRef]
83. Hecht, E. Energy and Change. *Phys. Teach.* **2007**, *45*, 88. [CrossRef]
84. Shapiro, S. *Thinking about Mathematics*; OUP: Oxford, UK, 2000.

85. Dinneen, J.D.; Brauner, C. Practical and Philosophical Considerations for Defining Information as Well-formed, Meaningful Data in the Information Sciences. *Library Trends* **2015**, *63*, 378–400. [CrossRef]

86. Heller, M. Dispute around sructural realism [Spór o realizm strukturalistyczny]. In *Filozofia i wszechświat: Wybór pism*; Krakow University: Krakow, Poland, 2006.

87. Venugopal, V. Scientists have turned the structure of the coronavirus into music. *Science* **2020**. [CrossRef]

88. Zhu, B.; Dacso, C.C.; O'Malley, B.W. Unveiling "Musica Universalis" of the Cell: A Brief History of Biological 12-Hour Rhythms. *J. Endocr. Soc.* **2018**, *2*, 727–752. [CrossRef]

89. The Music of the Spheres. BBC Radio 4 Discussion with Peter Forshaw, Jim Bennett & Angela Voss (In Our Time, June 19, 2008). Available online: https://www.bbc.co.uk/programmes/b00c1fct (accessed on 20 April 2020).

90. Snowdon, T.C.; Teie, D. Animal behaviour Affective responses in tamarins elicited by species-specific music. *Biol. Lett.* **2009**, *6*, 30–32. [CrossRef]

91. Shanahan, M. *The Technological Singularity*; MIT Press—Journals: Cambridge, MA, USA, 2015.

92. Dembski, W.A. Specification: The Pattern That Signifies Intelligence. 2005. Available online: https://billdembski.com/documents/2005.06.Specification.pdf (accessed on 12 November 2019).

93. Ney, A. Reductionism. Internet Encyclopedia Online. Available online: https://www.iep.utm.edu/red-ism/ (accessed on 20 April 2016).

94. van Riel, R.; van Gulick, R. *"Scientific Reduction", The Stanford Encyclopedia of Philosophy*. Edward, N., Ed.; Available online: https://plato.stanford.edu/entries/scientific-reduction/ (accessed on 20 April 2019).

95. Searle, J. *Minds, Brains and Science*; Harvard University Press: Cambridge, UK, 2003.

96. Searle, J.R. Why I am not a property dualist. *Philos. New Century* **2012**, *9*, 152–160. [CrossRef]

97. Searle, J.; Magazine, T.T.P. Interview—John Searle. *Philos. Mag.* **2008**, 55–58. [CrossRef]

98. Searle, J. Philosophy of Mind Lectures, Berkley Lecture 9. 2013. Available online: https://www.youtube.com/watch?v=9st8gGz0rOI (accessed on 1 March 2019).

99. Searle, J. Professor John Searle: Consciousness as a Problem in Philosophy and Neurobiology. [Online]. Available online: https://www.youtube.com/watch?v=6nTQnvGxEXw (accessed on 1 December 2019).

100. Dennett, D. *From Bacteria to Bach and Back: The Evolution of Minds*; W. W. Norton & Company: New York, NY, USA, 2017.

101. Millican, R. *Beyond Concepts*; Oxford University Press: Oxford, UK, 2017.

102. Piccinini, G.; Scarantino, A. Information Processing, Computation, and Cognition. *J.Biol. Phys.* **2011**, *37*, 1–38. [CrossRef]

103. James, S. American Poetry: The Nineteenth Century 99261 John Hollander Edited by. American Poetry: The Nineteenth Century. New York: Literary Classics of the United States 1993. 2 Vols. xxiii + 1099 pp.; xxxii + 1050 pp, ISBN: 1 57958 034 3, set distributed by Fitzroy Dearborn The Library of America series, Vols. 66–67 1998, distributed by Fitzroy-Dearborn, Chicago, and London. *Ref. Rev.* **1999**, *13*, 25–26. [CrossRef]

104. Hintikka, J. *Socratic Epistemology: Explorations of Knowledge—Seeking by Questioning*; Cambridge University Press: Cambridge, UK, 2007.

105. Kaplan, S. *The Experience of Nature*; Cambridge University Press: Cambridge, UK, 1989.

106. Bajić, V.B.; Tan, T.W. *Information Processing in Living Systems*; Imperial College Press: London, UK, 2005.

107. Baradian, X.; Moreno, A. On What Makes Certain Dynamical Systems Cognitive: A Minimal Cognitive Organization Program. Available online: https://philpapers.org/archive/BAROWM.pdf (accessed on 20 June 2020).

108. Yakura, H. A hypothesis: CRISPR-Cas as a minimal cognitive system. *Adapt. Behav.* **2019**, *27*. [CrossRef]

109. Jaworski, W. Hylemorphic Structure, Emergence and Supervenience. University of Oxford Podcasts—Audio and Video Lectures. 2018. Available online: https://podcasts.ox.ac.uk/hylomorphic-structure-emergence-and-supervenience (accessed on 1 August 2019).

110. Krzanowski, R. Towards a Formal Ontology of Information. Selected Ideas of K. Turek. *Philos. Probl. Sci.* **2016**, *61*, 23–52.

 *philosophies*

Article

Natural Morphological Computation as Foundation of Learning to Learn in Humans, Other Living Organisms, and Intelligent Machines

Gordana Dodig-Crnkovic [1,2]

1   Department of Computer Science and Engineering, Chalmers University of Technology and the University of Gothenburg, 40482 Gothenburg, Sweden; dodig@chalmers.se
2   School of Innovation, Design and Engineering, Mälardalen University, 721 23 Västerås, Sweden

Received: 3 July 2020; Accepted: 25 August 2020; Published: 1 September 2020

Abstract: The emerging contemporary natural philosophy provides a common ground for the integrative view of the natural, the artificial, and the human-social knowledge and practices. Learning process is central for acquiring, maintaining, and managing knowledge, both theoretical and practical. This paper explores the relationships between the present advances in understanding of learning in the sciences of the artificial (deep learning, robotics), natural sciences (neuroscience, cognitive science, biology), and philosophy (philosophy of computing, philosophy of mind, natural philosophy). The question is, what at this stage of the development the inspiration from nature, specifically its computational models such as info-computation through morphological computing, can contribute to machine learning and artificial intelligence, and how much on the other hand models and experiments in machine learning and robotics can motivate, justify, and inform research in computational cognitive science, neurosciences, and computing nature. We propose that one contribution can be understanding of the mechanisms of 'learning to learn', as a step towards deep learning with symbolic layer of computation/information processing in a framework linking connectionism with symbolism. As all natural systems possessing intelligence are cognitive systems, we describe the evolutionary arguments for the necessity of learning to learn for a system to reach human-level intelligence through evolution and development. The paper thus presents a contribution to the epistemology of the contemporary philosophy of nature.

Keywords: learning; learning to learn; deep learning; information processing; natural computing; morphological computing; info-computation; connectionism; symbolism; cognition; robotics; artificial intelligence

---

1. Introduction

Artificial intelligence in the form of machine learning is currently making impressive progress, especially in the field of deep learning (DL) [1]. Algorithms in deep learning have been inspired by the human brain, even though our knowledge about brain functions is still incomplete, yet steadily increasing. Learning here is a two-way process where computing is learning from neuroscience, while neuroscience is adopting information processing models, and this process iterates, as discussed in [2–4].

Deep learning is based on artificial neural networks resembling neural networks of the brain, processing huge amounts of (labelled) data by high-performance GPUs (graphical processing units) with a parallel architecture. It is (typically supervised) machine learning from examples. It is static, based on the assumption that the world behaves in a similar way and that domain of application is close to the domain where training data are obtained. However impressive and successful, deep-learning intelligence has an Achilles heel, and that is lack of common sense reasoning [5–7]. Its recognition of pictures is based on pixels, and small changes, even invisible for humans, can confuse deep learning algorithm, leading to very surprising errors. Bengio [5] therefore points out that deep learning is missing the capability of out-of-distribution generalization, and compositionality.

Human intelligence has two distinct mechanisms of learning according to Kahneman [8]—quick, bottom-up, from data to patterns (System 1) and slow, top-down from language to objects (System 2) which have been recognized and analyzed earlier [8–10]. The starting point of old AI (GOFAI) was System 2, symbolic, language, and logic-based reasoning, planning and decision making. However, it was without System 1, so it had a problem of symbol grounding, as its mappings were always in the space of representations from symbols to symbols, and never to the physical world itself.

Now deep learning has grounding for its symbols in the observed/collected/measured data, but it lacks the System 2 capabilities of generalization/symbol generation, symbol manipulation, and language that are necessary in order to get to the human-level intelligence and ability of learning and meta-learning, that is learning to learn. The step from (big) data-based System 1 to manipulation of few concepts like in high level reasoning of System 2 is suggested in [5] to proceed via concepts of agency, attention, and causality. It is expected that 'agent perspective' will help to put constraints on the learned representations and so to encapsulate causal variables, and affordances. Bengio proposes that "the agent perspective on representation learning should facilitate re-use of learned components in novel ways (..), enabling more powerful forms of compositional generalization, i.e., out-of-distribution generalization based on the hypothesis of localized (in time, space, and concept space) changes in the environment due to interventions of agents" [5].

This step, from System 1 (present state of DL) to System 2 (higher level cognition) will open new and even more powerful possibilities to AI. It is not the development into the unknown, as some of it on the System 2 side has earlier been proposed by GOFAI, and it is addressed in the new developments in cognitive science and neuroscience. In this article, we will focus on the modelling of System 1 and its connections to System 2 within the framework of computational model of cognition based on natural info-computation [3,4].

It should be acknowledged that the insight about the necessity of linking connectionism and symbolism is old, and already Minsky, in 1990, formulated the link in his "Logical vs. Analogical or Symbolic vs. Connectionist or Neat vs. Scruffy" [11]. For more recent reflections on the topic, see [12–14].

The article is structured as follows. After the introduction, learning about the world through agency is presented. Learning in the computing nature, including learning in the evolutionary perspective, is outlined in the subsequent section. We then address learning as computation in networks of agents, and info-computational learning by morphological computation. Learning to learn from raw data and up—agency from System 1 to System 2 is the last topic investigated. Conclusions and future work close the article.

2. Learning about the World through Agency

The framework for the discussion is the computing nature in the form of info-computationalism. It takes the world (*Umwelt*) for an agent to be information [15] with its dynamics seen as computation [16]. Information is observer relative and so is computation [17–19].

When discussing cognition as an embodied bioinformatic process, we use the notion of agent, i.e., a system able to act on its own behalf, pursuing an intrinsic goal [17,20]. Agency in biological systems, in the sense used here, has been explored in [21,22], where arguments are provided that the world as it appears to an agent depends on the type of agent and the type of interaction through which the agent acquires information about the world [17]. Agents communicate by exchanging messages (information), which helps them to coordinate their actions based on the information they possess and then they share through social cognition.

For something to be information, there must exist an agent for whom that is established as a "difference that makes a difference" [23]. When we argue that the fabric of the world is informational [24], the question can be asked: Who/what is the agent in that context? An agent can be seen as interacting with the points of inhomogeneity (differences that make a difference/data as atoms of information), establishing the connections between those data and the data that constitute the agent itself (a particle, a system). There are myriads of agents for which information of the world makes differences—from elementary particles to molecules, cells, organisms, societies … —all of them interact and exchange information on different levels of scale and these information dynamics are natural computation [25,26].

Our definition of agency and cognition as a property of all living organisms, is building on Maturana and Varela [27,28], and Stewart [29]. The question relevant for AI is how artifactual agents should be built in order to possess cognition and eventually even consciousness. Is it possible at all, given that cognition in living organisms is a deeply biologically rooted process? Along with reasoning, language is considered high-level cognitive activity that only humans are capable of. Increasing levels of cognition evolutionary developed in living organisms, starting from basic automatic behaviors such as found in organisms from bacteria [30–33] to insects, to increasingly complex behavior in complex multicellular life forms such as mammals [34]. Can AI "jump over" evolutionary steps in the development of cognition to reach, and even exceed, human intelligence?

While the idea that cognition is a biological process in all living organisms has been extensively discussed [27–29], it is not clear on which basis cognitive processes in all kinds of organisms would be accompanied by (some kind of, some degree of) consciousness. Consciousness is, according to Bengio [7], characteristics for System 2: "We closely associate conscious processing to Kahneman's system 2 cognitive abilities [Kahneman, 2011]." Bengio adopts Baars' global workspace theory of consciousness [35]. In the process of learning, and learning to learn, consciousness plays an important role through the process of attention, which selects only a tiny subset of information/data that is processed, instead of processing indiscriminately huge amounts of data, which is expensive from the point of view of response time and energy [7].

If we, in parallel with "minimal cognition" [36], search for "minimal consciousness" in an organism, what would that be? Opinions are divided at what point in the evolution one can say that consciousness emerged. Some would suggest as Liljenström and Århem that only humans possess consciousness, while the others are ready to recognize consciousness in animals with emotions [37,38]. From the info-computational point of view, it has been argued that cognitive agents with nervous systems are the step in evolution which first enabled consciousness in the sense of internal model with the ability of distinguishing the "self" from the "other" and provide representation of "reality" for an agent based on that distinction [4,39].

3. Learning in the Computing Nature

For naturalists, nature is the only reality [40]. Nature is described through its structures, processes, and relationships, using a scientific approach [41,42]. Naturalism studies the evolution of the entire natural world, including the life and development of human and humanity as a part of nature. Social and cultural phenomena are studied through their physical manifestations. An example of contemporary naturalist approach is the research field of social cognition with its network-based studies of social behaviors. Already Turing emphasized social character of learning [43], and was also elaborated on by Minsky [44] and Dennett [45].

Computational naturalism (pancomputationalism, naturalist computationalism, computing nature) [46–48], see even [3,4], is the view that the entirety of nature is a huge network of computational processes, which, according to physical laws, computes (dynamically develops) its own next state from the current one. Among prominent representatives of this approach are Zuse, Fredkin, Wolfram, Chaitin, and Lloyd, who proposed different varieties of computational naturalism. According to the idea of computing nature, one can view the time development (dynamics) of physical states as information processing (natural computation). Such processes include self-assembly, self-organization, developmental processes, gene regulation networks, gene assembly, protein–protein interaction networks, biological transport networks, social computing, evolution, and similar processes of morphogenesis (creation of form). The idea of computing nature and the relationships between two basic concepts of information and computation are explored in [17–19,25].

In the computing nature, cognition is a natural process, seen as a result of natural bio-chemical processes. All living organisms possess some degree of cognition, and for the simplest ones, like bacteria, cognition consists in metabolism and (my addition) locomotion [17]. This "degree" is not meant as continuous function, but as a qualitative characterization that cognitive capacities increase from simplest to the most complex organisms. The process of interaction with the environment causes changes in the informational structures that correspond to the body of an agent and its control mechanisms, which define its future interactions with the world and its inner information processing [49]. Informational structures of an agent become semantic information (i.e., get explicit metacognitive meaning through System 2, which generates metacognition for an agent) first in the case of highly intelligent agents capable of reasoning, which we know some birds are.

Recently, empirical studies have revealed an unexpected richness of cognitive behaviors (perception, information processing, memory, decision making) in organisms as simple as bacteria. [30–33]. Single bacteria are small, typically 0.5–5.0 micrometers in length, and interact with only their immediate environment. They live too short as a specific single organism to be able to memorize a significant amount of data. Biologically, bacteria are immortal at the level of the colony, as the two daughter bacteria from cell division of a parent bacterium are considered as two new individuals. Thus bacterial colonies, swarms, and films that extend to a bigger space, and can survive longer time, have longer memory, and exhibit an unanticipated complexity of behaviors that can undoubtedly be characterized as cognition [50,51], see even [45]. More fascinating cases are even simpler agents like viruses, on the border of the living, which are based on the simple principle that the most viable versions persist and multiply while others vanish [52,53]. Memory and learning are the key competences of living organisms [50], and in the simplest case, memory is based on the change of shape [54], which appears on different scales and levels of organization [55]. Fields and Levin add evolutionary perspective to the memory characterization and argue that "genome is only one of several multi-generational biological memories". Additionally, cytoplasm and cell membrane, which characterize all of life on evolutionary timescale, preserve memory [56]. Because of complex structure of the cell, biological memory cannot be understood at one particular scale, and information is propagated and preserved in non-genomic cellular structures, which changes current understanding of biological memory [55,56]. It also forms at different time scales [57].

Starting with bacteria and archaea [58], all organisms without nervous systems cognize, that is perceive their environment, process information, learn, memorize, and communicate. As they are natural information processors, some such as slime mold, multinucleate, or multicellular Amoebozoa, has been used as natural computer/information processors to compute shortest paths. Even plants cognize, in spite of being often thought of as living systems without cognitive capacities [59]. Plants have been found to possess memory (in their bodily structures that change as a result of past events), the ability to learn (plasticity, ability to adapt through morphodynamics), and the capacity to anticipate and direct their behavior accordingly. Plants are also argued to possess rudimentary forms of knowledge, according to [60] (p. 121), [61] (p. 7), and [34] (p. 61).

Consequently, in this article we take basic cognition to be the totality of processes of self-generation/self-organization, self-regulation, and self-maintenance that enables organisms to survive processing information from the environment. The understanding of cognition as it appears in degrees of complexity in living nature can help us better understand the step between inanimate and animate matter from the first autocatalytic chemical reactions to the first autopoietic proto-cells, as well as evolution of life and learning.

Learning in the Evolutionary Perspective

A recent trend in the design is learning from nature, biomimetics. Deep learning is one of the technologies developed within the biomimetic paradigm. In the case of intelligence, we still have a lot to learn from nature about how our own brains, intelligence, and learning function. One of the strategies is to start with learning in simplest organisms, in order to uncover basic mechanisms of the process. Evolution can be seen as a process of problem-solving [34]. "From the amoeba to Einstein, the growth of knowledge is always the same: We try to solve our problems, and to obtain, by a process of elimination something approaching adequacy in our tentative solutions" [62] (p. 261). All acquired knowledge—whether it is acquired in the process of genetic evolution or in the process of individual learning—consists, this is Popper's central claim, in the modification "of some form of knowledge, or disposition, which was there previously, and in the last instance of inborn expectations" [62] (p. 71).

Popper's theory of the growth of knowledge through trial-and-error conjecture-based problem-solving shares basic approach with evolutionary epistemology. According to Campbell [63], all knowledge processes can be seen as the "variation and selective retention process of evolutionary adaptation" [64]. Thagard [65] criticizes Popper, Campbell, Toulmin, and others who proposed Darwinian models of the growth of (scientific) knowledge. Evolutionary epistemology emphasizes analogy between the development of biological species and scientific knowledge, based on variation, selection, and transmission. Thagard, on the other hand, holds that differences are more important than similarities, and that scientific knowledge is guided by "intentional, abductive theory construction in scientific discovery, the selection of theories according to general criteria, the achievement of progress by sustained application of criteria, and the transmission of selected theories in highly organized scientific communities". Even though scientific knowledge is a specific, formal kind of knowledge, it is nevertheless knowledge.

This criticism of evolutionary epistemology is addressing a specific understanding of evolution, through Darwinism in the narrow sense. However, the contemporary extended evolutionary synthesis provides mechanisms beyond blind variation of narrow Darwinism, and can accommodate for learning, anticipation, and intentionality [66–69]. In a similar, broader evolutionary approach, Watson and Szathmáry ask "Can evolution learn?" in [70], and suggest that "evolution can learn in more sophisticated ways than previously realized". Here "A system exhibits learning if its performance at some task improves with experience". They propose new theoretical approaches to the evolution of evolvability, and the evolution of ecological organizations, among others. They refer to Turing, who made an algorithmic model of computation (Turing machine) and established the connection between learning and intelligence through an algorithmic approach [71]. The relationship between learning and evolution is established through the notion of reinforcement learning, as "reusing behaviors that have been successful in the past (reinforcement learning) is intuitively similar to the way selection increases the proportion of fit phenotypes in a population". Watson and Szathmáry's paper list number of different types of learning, including diverse machine learning approaches, ended with the claim that there is a clear analogy between evolution and the process of learning, and that we can better understand evolution if we see it as learning.

In spite of mentioning Turing's pioneering work on the topic of algorithmic learning, Watson and Szathmáry assume "incremental adaptation (e.g., from positive and/or negative reinforcement)".

Critics of the evolutionary approach argue for the impossibility of such incremental process to produce highly complex structures such as intelligent living organisms. Monkeys typing Shakespeare are often used as illustration. As an counterargument, Chaitin [72] pointed out that typing monkeys' argument does not take into account physical laws of the universe, which dramatically limit what can be typed. Moreover, the universe is not a typewriter, but a computer, so a monkey types random input into a computer. The computer interprets the strings as programs. Or, in the words of Gershenfeld: "Your genome doesn't store anywhere that you have five fingers. It stores a developmental program, and when you run it, you get five fingers" [73].

Sloman argued that "many of the developments in biological evolution that are so far not understood, and in some cases have gone unnoticed, were concerned with changes in information processing. The same is true of changes in individual development and learning: They often produce new forms of information processing". He addressed this phenomenon through computational ideas about morphogenesis and meta-morphogenesis [74]. His approach offers new insight, that variation is algorithmic. To Sloman's computational approach, I would add that steps in variation are morphological computation, which means physical computation, capable of randomly modifying genes, and executing morphological programs which do not present smooth incremental changes, but considerable jumps in properties of structures and processes. Morphological computation acts also through gene regulation, which is one more process that was unknown to both Darwin and to proponents of evolution as Modern Synthesis. Originally, genes were considered as coding for specific proteins, and it was believed that all genes were active. Gene regulation involves a mechanism that can repress or induce the expression of a gene. According to Nature [75], "These include structural and chemical changes to the genetic material, binding of proteins to specific DNA elements to regulate transcription, or mechanisms that modulate translation of mRNA".

4. Learning as Computation in Networks of Agents

In what follows, we will focus on info-computational framework of learning. Informational structures constituting the fabric of physical nature are networks of networks, which represent semantic relations between data for an agent [18]. Information is organized in levels or layers, from quantum level to atomic, molecular, cellular, organismic, social, and so on. Computation/information processing involves data structure exchanges within informational networks, which are instructively represented by Carl Hewitt's actor model [76]. Different types of computation emerge at different levels of organization in nature as exchanges of informational structures between the nodes (computational agents) in the network [17].

The research in computing nature/natural computing is characterized by bi-directional knowledge exchanges, through the interactions between computing and natural sciences [54]. While natural sciences are adopting tools, methodologies, and ideas of information processing, computing is broadening the notion of computation, taking information processing found in nature as computation [2,77]. Based on that, Denning argues that computing today is a natural science, the fourth great domain of science [78,79]. Computation found in nature is a physical process, where nature computes with physical bodies as objects. Physical laws govern processes of computation, which appear on many different levels of organization in nature.

With its layered computational architecture, natural computation provides a basis for a unified understanding of phenomena of embodied cognition, intelligence, and learning (knowledge generation), including meta-learning (learning to learn) [47,80]. Natural computation can be modelled as a process of exchange of information in a network of informational agents [76], i.e., entities capable of acting on their own behalf, which is Hewitt's actor model applied to natural agents.

One sort of computation is found on the quantum-mechanical level, where agents are elementary particles, and messages (information carriers) are exchanged by force carriers, while different types of computation can be found on other levels of organization in nature. In biology, information processing is going on in cells, tissues, organs, organisms, and eco-systems, with corresponding agents and message types. In biological computing, the message carriers are chunks of information such as molecules, while in social computing, they are sentences while the computational nodes (agents) are molecules, cells, and organisms in biological computing or groups/societies in social computing [19].

5. Info-Computational Learning by Morphological Computation

The notion of computation in this framework refers to the most general concept of intrinsic computation, that is spontaneous computation processes in the nature [2,77], and which is used as a basis of designed computation found in computing machinery [81]. Intrinsic natural computation includes quantum computation [81,82], processes of self-organization, self-assembly, developmental processes, gene regulation networks, gene assembly, protein–protein interaction networks, biological transport networks, and similar. It is both analog (such as found in dynamic systems) and digital.

The majority of info-computational processes are sub-symbolic and some of them are symbolic (like reasoning and languages).

Within info-computational framework, or computing nature [18], computation on a given level of organization of information presents a realization/actualization of the laws that govern interactions between its constituent parts. On the basic level, computation is manifestation of causation in the physical substrate [83]. In every next layer of organization, a set of rules governing the system switch to the new emergent regime. It remains yet to be established how this process exactly goes on in nature, and how emergent properties occur [84]. Research on natural computing is expected to uncover those mechanisms. In the words of Rozenberg and Kari: "(O)ur task is nothing less than to discover a new, broader, notion of computation, and to understand the world around us in terms of information processing" [2]. From the research in complex dynamical systems, biology, neuroscience, cognitive science, networks, concurrency etc., new insights essential for the info-computational nature are steadily coming. Here it should be mentioned that the computing nature with "bold" physical computation [85] is the maximal physicalist approach to computing. There are less radical approaches, such as taken by Horsman, Stepney, and co-authors [86–88], known as Abstraction/Representation theory (AR theory), where "physical computing is the use of a physical system to predict the outcome of an abstract evolution", where computation defines the relationship between physical systems and abstract concepts/representations. Unlike AR theory, info-computationalism also embraces computation without representation, in the sense of Brooks [89] or Pfeifer [90]. Even though it is already established that the original Turing model of computation is specific and represents a human performing calculation, as pointed out by Copeland [91], even Turing himself started exploring computation beyond the Turing Machine model.

Turing's 1952 paper [92] may be considered as a predecessor of natural computing. It addressed the process of morphogenesis by proposing a chemical model as the explanation of the development of biological patterns such as the spots and stripes on animal skin. Turing did not claim that a physical system producing patterns actually performed computation. From the perspective of computing nature, we can now argue that morphogenesis is a process of morphological computation. Informational structure (as representation of embodied physical structure) presents a program that governs computational process [93], which in its turn changes that original informational structure following/implementing/realizing physical laws.

Morphology is the central idea in our understanding of the connection between computation and information. Morphological/morphogenetic computing on that informational structure leads to new informational structures via processes of self-organization of information. Evolution itself is a process of morphological computation on a long-term scale. It is also important to take into account the second order process of morphogenesis of morphogenesis (meta-morphogenesis) as done by Sloman [74].

A closely related idea to natural computing is Valiant's [94] view of evolution by "ecorithms"—learning algorithms that perform "probably approximately correct" (PAC) computation. Unlike the classical model of Turing machine, the "ecorithmic" computation does not give perfect results, but good enough (for an agent). That is the case for natural computing in biological agents who always act under resource constraints, especially time, energy, and material limitations, unlike Turing machine model of computation that by definition operates with unlimited resources. An older term for PAC due to Simon is "satisficing" [95] (p. 129): "Evidently, organisms adapt well enough to 'satisfice'; they do not, in general, 'optimize'".

6. Learning to Learn from Raw Data and up—Agency from System 1 to System 2

Cognition is a result of a processes of morphological computation on informational structures of a cognitive agent in the interaction with the physical world, with processes going on at both sub-symbolic and symbolic levels [4]. This morphological computation establishes connections between an agent's body, its nervous system (control), and its environment [49]. Through the embodied interaction with the informational structures of the environment, via sensory-motor coordination, information structures are induced (stimulated, produced) in the sensory data of a cognitive agent, thus establishing perception, categorization, and learning. Those processes result in constant updates of memory and other structures that support behavior, particularly anticipation. Embodied and corresponding induced informational structures (in the Sloman's sense of virtual machine) [96] are the basis of all cognitive activities, including consciousness and language as a means of maintenance of "reality" or the representation of the world in the agent.

From the simplest cognizing agents such as bacteria to the complex biological organisms with nervous systems and brains, the basic informational structures undergo transformations through morphological computation as developmental and evolutionary form-generation—morphogenesis. Living organisms as complex agents inherit bodily structures resulting from a long evolutionary development of species. Those structures are the embodied memory of the evolutionary past [54]. They present the means for agents to interact with the world, get new information that induces embodied memories, learn new patterns of behavior, and learn/construct new knowledge. By Hebbian learning in the brain (where neurons that wire together, fire together, and habits increase probability of firing), world shapes humans' (or an animals') informational structures. Neural networks that "self-organize stable pattern recognition code in real-time in response to arbitrary sequences of input patterns" are an illustrative example [97].

On the fundamental level of quantum mechanical substrate, information processes represent actions of laws of physics. Physicists are already working on reformulating physics in terms of information [98–103]. This development can be related to the Wheeler's idea "it from bit" [104] and von Weizsäcker's ur-alternatives [105].

In the computing nature approach, nature is consisting of physical structures that form levels of organization, on which computation processes develop. It has been argued that on the lower levels of organization, finite automata or Turing machines might be an adequate model of computation, while in the case of human cognition on the level of the whole-brain, non-Turing computation is necessary, see Ehresmann [106] and Ghosh et al. [107]. Symbols on the higher levels of abstraction (System 2) are related with several possible sub-symbolic realizations, which they generalize, as Ehresmann's models show. Zenil et al.'s work on causality by algorithmic generative models to "decompose an observation into its most likely algorithmic generative models" [108] presents one of the recent attempts to computationally/algorithmically approach causality. Algorithmic computation is a very important part of computational models defined by Turing, based on symbol manipulation. The connection to sub-symbolic is done through algorithmic information theory.

Apart from the *Handbook of Natural Computing* [77] that presents concrete models of natural computation, interesting work on computational modelling of biochemistry and reaction networks have been done by Cardelli [109–112], including the study of morphisms of reaction networks that link structure to function. On the side of cognitive computing, Fresco addresses the physical computation and its role in cognition [113].

Principles of morphological computing and data self-organization from biology have been applied in robotics as well. In recent years, morphological computing emerged as a new idea in robotics, see [3,4] and references therein. Initially, robotics treated separately the body as a machine, and its control as a program. Meanwhile it has become apparent that embodiment itself is fundamental for cognition, generation of behavior, intelligence, and learning. Embodiment is central because cognition arises from the interaction of brain, body, and environment [90]. Agents' behavior develops through embodied interaction with the environment, in particular through sensory-motor coordination, when information structure is induced in the sensory data, thus leading to perception, learning, and categorization [48]. Morphological computing has also been applied in soft robotics, self-assembly systems, and molecular robotics, embodied robotics, and more. Even though the use of morphological computing in robotics is slightly different from the one in computing nature, there are common grounds and possibilities to learn from each other on the multidisciplinary level. Similar goes for the research being done in the fields of cognitive informatics and cognitive computing. There are also important connections to computational mechanics, algorithmic information dynamics (probabilistic framework of algorithmic information dynamics used for causal analysis), and neuro-symbolic computation, combining symbolic and neural processing, all of which are in different ways relevant to the topic. Those connections remain to explore in the future work.

7. Conclusions and Future Work

The info-computational approach, developed by the author, with natural morphological computation as a basis, is used to approach learning and learning to learn in humans, other living organisms, and intelligent machines. This paper is a contribution to epistemology of the philosophy of nature, proposing a new perspective on the learning process, both in artificial information processing systems such as robots and AI systems, and in natural information processing systems like living organisms.

Morphological computation is proposed as a mechanism of learning and meta-learning, necessary for connecting the pre-symbolic (pre-conscious) with the symbolic (conscious) information processing. In the framework of info-computational nature, morphological computation is information (re)structuring through computational processes which follow (implement) physical laws. It is grounded in the notion of agency, with causality represented by morphological computation.

Morphology is the central idea in understanding of the connection between computation (morphological/morphogenetical) and information. Morphology refers to form, shape, and structure. Materials represent morphology on the underlying level of organization. For the arrangements of molecular and atomic structures, material are protons, neutrons, and electrons on the level below.

Morphological computation, represented as information communication between agents/actors of the Hewitt actor model, distributed in space, where computational devices communicate asynchronously and the entire computation is generally not in any well-defined state [3]. Unlike Turing computation, which is a mathematical–logical model, Hewitt computation is a physical model. For morphological computing as information (re)structuring through computational processes which follow (implement) physical laws, Hewitt computation provides consequent formalization. On the basic level, morphological computation is natural computation in which physical objects perform computation. Symbol manipulation in this case is physical object manipulation, in the sense of Brooks "the world is its own best model". It becomes relevant in robotics and deep learning that manage direct behavior of an agent in the physical world.

In morphological computation, cognition is the restructuring of an agent through the interaction with the world, so all living organisms possess some degree of cognition. As a result of evolution, increasingly complex living organisms arise from the simple ones, that are able to survive and adapt to their environment. It means they are able to register inputs (data) from the environment, to structure those into information, and in more developed organisms into knowledge. The evolutionary advantage of using structured, component-based approaches is improving response-time and efficiency of cognitive processes of an organism, which drives the development from organisms with learning on the System 1 level, to the ones that acquire System 2 capabilities on top of it. In more complex cognitive agents, knowledge is built upon not only reaction to input information, but also on internal information processing with intentional choices, dependent on value systems stored and organized in agents' memory.

Knowledge generation places information and computation (communication) in focus, as information and its processing are essential structural and dynamic elements which characterize structuring of input data (data → information → knowledge → metaknowledge) by an interactive computational process going on in the agent during the adaptive interplay with the environment.

In nature, through the process of evolution and development, living systems learn to survive and thrive in their environment. Interactions present forms of reinforcement learning or Hebbian learning that make previous successful strategies preferred in the future [70]. That happens on a variety of levels of organization. On the meta-level, the meta-morphological computing (as a Sloman's virtual machine) [96] governs learning to learn.

In the case of human learning, the brain as a network of computational agents processes information obtained through the embodied communication with the environment as well as internal information from the body. Consciousness is a process of integration of information in the brain [35], and it gets a huge amount of data/information that would be overwhelming for the brain to handle in real time, so it uses the mechanism of attention to focus on a specific subset of information, typically regarding agent-based processes in the world. There changes in the scene are the consequence of the agent's interactions, and they are the unfolding of physical processes of morphological computations. Causality, or rather stable correlations between structures and processes in the world (from an agent's perspective) follow from what humans learn/memorize, as they get organized internally through the Hebbian principles where neurons that fire together, wire together.

Sloman, who developed theory of Meta-morphogenesis [74], started with the idea that changes in individual development and learning of an agent produce new forms of information processing [74]. His approach offers new insight, that variation is algorithmic. The interplay between structure and process is essential for learning, as past experiences stored in structures affect the possibility of future processes and strategies of learning and learning to learn. To Sloman's morphogenetic approach, I would propose to add that steps in variation are results of morphological computation, which means physical computation, capable of, e.g., modifying genes, and executing morphological programs which do not present smooth incremental changes, but jumps in properties of structures and processes. Morphological computation acts also through gene regulation, which is one more process that was unknown to both Darwin and to proponents of evolution as Modern Synthesis.

Since contemporary deep-learning-centered AI (dealing with human-level cognition and above) is gradually developing from the present state System 1 (connectionist, sub-symbolic) coverage towards the System 2 (symbolic), with agency, causality, consciousness, and attention as mechanisms of learning and meta-learning [5,114], it searches for mechanisms of transition between two systems. An inspiration for technology development, the human brain is of interest as the center of learning in humans, that is self-organized, resilient, fault tolerant, plastic, computationally powerful, and energetically efficient. In its development, like in the past, deep learning is inspired by nature, assimilating ideas from neuroscience, cognitive science, biology, and more. The AI approach to understanding, via decomposition and construction, is close to other computational models of nature in that it seeks testable and applicable models, based on data and information processing. Bengio's proposal of agent-based perspective [5], necessary to proceed from System 1 to System 2 learning, can be related to the model of learning based on morphological computing.

For the future, more interdisciplinary/crossdisciplinary/transdisciplinary work remains to be done as a way to increase understanding of connections between the low level and the high level cognitive processes, learning, and meta-learning. It will also be instructive to find relations between (levels of/degrees of) cognition and consciousness as mechanisms helping to reduce the number of variables that are manipulated by an agent for the purpose of perception, reasoning, decision-making, planning acting/agency, and learning.

The goals of artificial intelligence, as well as robotics, differ from those of the computing nature and morphological computing. AI builds solutions for practical problems and in that it typically focuses on the highest possible level of intelligence, even though among the AI fields inspired by computing nature, there is developmental robotics, which has more explorative character.

The priority of info-computational naturalism is understanding and connecting knowledge about nature, while a lot of current technology is searching for inspiration in nature in pursuit of new technological solutions. Paths of the two are meeting, and mutual exchange of ideas is beneficial for both sides. Specialist sciences and philosophies also need close communication and exchange of ideas. Learning and meta-learning within computing nature is such a topic of central importance that calls for more knowledge from a variety of fields. This paper is not only the presentation of how much that is already known, but also an attempt to indicate how much more remains to be done.

Funding: This research is funded by Swedish Research Council, VR grant MORCOM@COG.

Acknowledgments: The author would like to thank anonymous reviewers for very helpful, constructive, and instructive review comments.

Conflicts of Interest: The author declares no conflict of interest. The funders had no role in the design of the study, in the writing of the manuscript, or in the decision to publish the results.

References

1. Lecun, Y.; Bengio, Y.; Hinton, G. Deep Learning. *Nature* **2015**, *521*, 436–444. [CrossRef] [PubMed]
2. Rozenberg, G.; Kari, L. The Many Facets of Natural Computing. *Commun. ACM* **2008**, *51*, 72–83.
3. Dodig-Crnkovic, G. Nature as a network of morphological infocomputational processes for cognitive agents. *Eur. Phys. J. Spec. Top.* **2017**. [CrossRef]
4. Dodig-Crnkovic, G. Cognition as Embodied Morphological Computation. In *Philosophy and Theory of Artificial Intelligence*; Studies in Applied Philosophy, Epistemology and Rational Ethics; Springer: Cham, Switzerland, 2018; Volume 44, pp. 19–23.
5. Bengio, Y. From System 1 Deep Learning to System 2 Deep Learning (NeurIPS 2019). Available online: https://www.youtube.com/watch?v=T3sxeTgT4qc (accessed on 24 June 2020).
6. Bengio, Y. Scaling up deep learning. In *Proceedings of the KDD '14. Proceedings of the 20th ACM SIGKDD International Conference on Knowledge Discovery and Data Mining*; ACM: New York, NY, USA, 2014; p. 1966.
7. Bengio, Y. The Consciousness Prior. *arXiv* **2019**, arXiv:1709.08568v2.
8. Kahneman, D. *Thinking, Fast and Slow*; Farrar, Straus and Giroux: New York, NY, USA, 2011; ISBN 9780374275631.
9. Clark, A. *Microcognition: Philosophy, Cognitive Science, and Parallel Distributed Processing*; MIT Press: Cambridge, MA, USA, 1989; ISBN 978-0262530958.
10. Scellier, B.; Bengio, Y. Towards a Biologically Plausible Backprop. *arXiv* **2016**, arXiv:1602.05179, 1–17.
11. Minsky, M. Logical vs. Analogical or Symbolic vs. Connectionist or Neat vs. Scruffy. In *Artificial Intelligence at MIT, Expanding Frontiers*; Winston, P.H., Ed.; MIT Press: Cambridge, MA, USA, 1990.
12. Dinsmore, J. *The Symbolic and Connectionist Paradigms*; Psychology Press: New York, NY, USA; London, UK, 2014; ISBN 978-0805810806.
13. Wang, J. Symbolism vs. Connectionism: A Closing Gap in Artificial Intelligence. Available online: http://wangjieshu.com/2017/12/23/symbol-vs-connectionism-a-closing-gap-in-artificial-intelligence/ (accessed on 28 June 2020).
14. Garcez, A.D.A.; Besold, T.R.; De Raedt, L.; Foldiak, P.; Hitzler, P.; Icard, T.; Kiihnberger, K.U.; Lamb, L.C.; Miikkulainen, R.; Silver, D.L. Neural-symbolic learning and reasoning: Contributions and challenges. In Proceedings of the AAAI Spring Symposium-Technical Report, Stanford, CA, USA, 23–25 March 2015; Dagstuhl Seminar 14381. Dagstuhl Publishing: Dagstuhl, Germany, 2015.
15. Floridi, L. Informational realism. In *Proceedings of the Selected Papers from Conference on Computers and Philosophy-Volume 37 (CRPIT '03)*; Weckert, J., Al-Saggaf, Y., Eds.; Australian Computer Society, Inc.: Darlinghurst, Australia, 2003; pp. 7–12.
16. Dodig-Crnkovic, G. Dynamics of Information as Natural Computation. *Information* **2011**, *2*, 460–477. [CrossRef]
17. Dodig-Crnkovic, G. Information, Computation, Cognition. Agency-Based Hierarchies of Levels. In *Fundamental Issues of Artificial Intelligence. Synthese Library, (Studies in Epistemology, Logic, Methodology, and Philosophy of Science)*; Müller, V., Ed.; Springer International Publishing: Cham, Switzerland, 2016; pp. 141–159. ISBN 9783319264851. Volume 376.
18. Dodig-Crnkovic, G.; Giovagnoli, R. *COMPUTING NATURE. Turing Centenary Perspective*; Dodig-Crnkovic, G., Giovagnoli, R., Eds.; Springer: Berlin/Heidelberg, Germany, 2013; Volume 7, ISBN 978-3-642-37224-7.
19. Dodig-Crnkovic, G. Physical computation as dynamics of form that glues everything together. *Information* **2012**, *3*, 204–218. [CrossRef]
20. Froese, T.; Ziemke, T. Enactive Artificial Intelligence: Investigating the systemic organization of life and mind. *Artif. Intell.* **2009**, *173*, 466–500. [CrossRef]

21. Kauffman, S. *The Origins of Order: Self-Organization and Selection in Evolution*; Oxford University Press: Oxford, UK, 1993; ISBN 978-0195079517.

22. Deacon, T. *Incomplete Nature. How Mind Emerged from Matter*; W. W. Norton & Company: New York, NY, USA, 2011; ISBN 978-0-393-04991-6.

23. Bateson, G. *Steps to an Ecology of Mind*; University of Chicago Press: Chicago, IL, USA, 1973; ISBN 9780226039053.

24. Floridi, L. A defense of informational structural realism. *Synthese* **2008**, *161*, 219–253. [CrossRef]

25. Chaitin, G. Building the World Out of Information and Computation: Is God a Programmer, Not a Mathematician? In *Exploring the Foundations of Science, Thought and Reality*; Wuppuluri, S., Doria, F.A., Eds.; Springer International Publishing: Cham, Switzerland, 2018.

26. Dodig-Crnkovic, G. Shifting the Paradigm of Philosophy of Science: Philosophy of Information and a New Renaissance. *Minds Mach.* **2003**. [CrossRef]

27. Maturana, H. *Biology of Cognition*; Defense Technical Information Center: Urbana, IL, USA, 1970.

28. Maturana, H.; Varela, F. *Autopoiesis and cognition: The Realization of the Living*; D. Reidel Publishing Company: Dordrecht, The Netherlands, 1980; ISBN 9789027710154.

29. Stewart, J. Cognition = life: Implications for higher-level cognition. *Behav. Process.* **1996**, *35*, 311–326. [CrossRef]

30. Ben-Jacob, E. Bacterial Complexity: More Is Different on All Levels. In *Systems Biology-The Challenge of Complexity*; Nakanishi, S., Kageyama, R., Watanabe, D., Eds.; Springer: Berlin/Heidelberg, Germany, 2009; pp. 25–35.

31. Ben-Jacob, E. Learning from Bacteria about Natural Information Processing. *Ann. N. Y. Acad. Sci.* **2009**, *1178*, 78–90. [CrossRef] [PubMed]

32. Lyon, P. The cognitive cell: Bacterial behavior reconsidered. *Front. Microbiol.* **2015**, *6*, 264. [CrossRef] [PubMed]

33. Marijuán, P.C.; Navarro, J.; del Moral, R. On prokaryotic intelligence: Strategies for sensing the environment. *BioSystems* **2010**. [CrossRef]

34. Popper, K. *All Life Is Problem Solving*; Routledge: London, UK, 1999; ISBN 978-0415249928.

35. Baars, B.J. Global workspace theory of consciousness: Toward a cognitive neuroscience of human experience. *Prog. Brain Res.* **2005**, *150*, 45–53. [CrossRef]

36. Van Duijn, M.; Keijzer, F.; Franken, D. Principles of Minimal Cognition: Casting Cognition as Sensorimotor Coordination. *Adapt. Behav.* **2006**, *14*, 157–170. [CrossRef]

37. Århem, P.; Liljenström, H. On the coevolution of cognition and consciousness. *J. Theor. Biol.* **1997**. [CrossRef]

38. Liljenström, H.; Århem, P. *Consciousness Transitions: Phylogenetic, Ontogenetic and Physiological Aspects*; Elsevier: Amsterdam, The Netherlands, 2011; ISBN 978-0-444-52977-0.

39. Dodig-Crnkovic, G.; von Haugwitz, R. Reality Construction in Cognitive Agents through Processes of Info-Computation. In *Representation and Reality in Humans, Other Living Organisms and Intelligent Machines*; Dodig-Crnkovic, G., Giovagnoli, R., Eds.; Springer International Publishing: Basel, Switzerland, 2017; pp. 211–235. ISBN 978-3-319-43782-8.

40. Putnam, H. *Mathematics, Matter and Method*; Cambridge University Press: Cambridge, MA, USA, 1975.

41. Dodig-Crnkovic, G.; Schroeder, M. Contemporary Natural Philosophy and Philosophies. *Philosophies* **2018**, *3*, 42. [CrossRef]

42. Dodig-Crnkovic, G.; Schroeder, M. *Contemporary Natural Philosophy and Philosophies-Part 1*; MDPI AG: Basel, Switzerland, 2019; ISBN 978-3-03897-822-0.

43. Edmonds, B.; Gershenson, C. Learning, Social Intelligence and the Turing Test. In *How the World Computes. CiE 2012. Lecture Notes in Computer Science*; Cooper, S.B., Dawar, A., Löwe, B., Eds.; Springer: Berlin/Heidelberg, Germany, 2012; Volume 7318.

44. Minsky, M. *The Society of Mind*; Simon and Schuster: New York, NY, USA, 1986; ISBN 0-671-60740-5.

45. Dennett, D. *From Bacteria to Bach and Back: The Evolution of Minds*; W. W. Norton & Company: New York City, NY, USA, 2017; ISBN 978-0-393-24207-2.

46. Dodig-Crnkovic, G. *Investigations into Information Semantics and Ethics of Computing*; Mälardalen University Press: Västerås, Sweden, 2006; ISBN 91-85485-23-3.

47. Dodig-Crnkovic, G.; Müller, V. A Dialogue Concerning Two World Systems: Info-Computational vs. Mechanistic. In *Information and Computation*; Dodig Crnkovic, G., Burgin, M., Eds.; World Scientific Pub Co Inc: Singapore, 2009; pp. 149–184. ISBN 978-981-4295-47-5.

48. Dodig-Crnkovic, G. The info-computational nature of morphological computing. In *Philosophy and Theory of Artificial Intelligence. Studies in Applied Philosophy, Epistemology and Rational Ethics*; Müller, V., Ed.; Springer: Berlin/Heidelberg, Germany, 2013; Volume 5, ISBN 00935301.

49. Pfeifer, R.; Bongard, J. *How the Body Shapes the Way We Think–A New View of Intelligence*; MIT Press: Cambridge, MA, USA, 2006; ISBN 9780262162395.

50. Witzany, G. Memory and Learning as Key Competences of Living Organisms. In *Memory and Learning in Plants. Signaling and Communication in Plants*; Baluska, F., Gagliano, M., Witzany, G., Eds.; Springer Nature Switzerland: Cham, Switzerland, 2018; pp. 1–16.

51. Witzany, G. Introduction: Key Levels of Biocommunication of Bacteria. In *Biocommunication in Soil Microorganisms*; Witzany, G., Ed.; Springer: Berlin/Heidelberg, Germany, 2011.

52. Witzany, G. *Viruses: Essential Agents of Life*; Springer Netherlands: Dodrecht, Netherlands, 2012; ISBN 9789400748996.

53. Villarreal, L.P.; Witzany, G. Viruses are essential agents within the roots and stem of the tree of life. *J. Theor. Biol.* **2010**. [CrossRef]

54. Leyton, M. Shape as Memory Storage. In *Ambient Intelligence for Scientific Discovery. Lecture Notes in Computer Science*; Yang, C., Ed.; Springer: Berlin/Heidelberg, Germany, 2005; Volume 3345.

55. Kandel, E.R.; Dudai, Y.; Mayford, M.R. The molecular and systems biology of memory. *Cell* **2014**. [CrossRef]

56. Fields, C.; Levin, M. Multiscale memory and bioelectric error correction in the cytoplasm–cytoskeleton–membrane system. *Wiley Interdiscip. Rev. Syst. Biol. Med.* **2018**. [CrossRef] [PubMed]

57. Kukushkin, N.V.; Carew, T.J. Memory Takes Time. *Neuron* **2017**. [CrossRef] [PubMed]

58. Witzany, G. *Biocommunication of Archaea*; Springer International Publishing: Cham, Switzerland, 2017; ISBN 9783319655369.

59. Witzany, G. Bio-communication of Plants. *Nat. Preced.* **2007**. [CrossRef]

60. Pombo, O.; Torres, J.M.; Rahman, S. *Special Sciences and the Unity of Science*; Logic, E., Ed.; Springer: Berlin/Heidelberg, Germany, 2012; ISBN 978-94-007-9213-5.

61. Rosen, R. *Anticipatory Systems*; Springer: New York, NY, USA, 1985; ISBN 978-1-4614-1268-7.

62. Popper, K. *Objective Knowledge: An Evolutionary Approach*; Oxford University Press: Oxford, UK, 1972.

63. Campbell, D.T. Evolutionary epistemology. In *The Philosophy of Karl Popper*; Schilpp, P.A., Ed.; Open Court Publ.: La Salle, IL, USA, 1974; Volume 1, pp. 413–463.

64. Vanberg, V. Cultural Evolution, Collective Learning, and Constitutional Design. In *Economic Thought and Political Theory*; Reisman, D., Ed.; Springer: Dordrecht, The Netherlands, 1994. [CrossRef]

65. Thagard, P. Against Evolutionary Epistemology. *PSA Proc. Bienn. Meet. Philos. Sci. Assoc.* **1980**. [CrossRef]

66. Kronfeldner, M.E. Darwinian "blind" hypothesis formation revisited. *Synthese* **2010**. [CrossRef]

67. Jablonka, E.; Lamb, M.J.; Anna, Z. *Evolution in Four Dimensions: Genetic, Epigenetic, Behavioral, and Symbolic Variation in the History of Life*; MIT Press: Cambridge, MA, USA, 2014; ISBN 9780262322676.

68. Laland, K.N.; Uller, T.; Feldman, M.W.; Sterelny, K.; Müller, G.B.; Moczek, A.; Jablonka, E.; Odling-Smee, J. The extended evolutionary synthesis: Its structure, assumptions and predictions. *Proc. R. Soc. B Biol. Sci.* **2015**. [CrossRef] [PubMed]

69. Noble, D. The Music of Life: Biology Beyond the Genome. *Lavoisierfr* **2006**. [CrossRef]

70. Watson, R.A.; Szathmáry, E. How Can Evolution Learn? *Trends Ecol Evol.* **2016**, *31*, 147–157. [CrossRef]

71. Turing, A.M. Computing machinery and intelligence. In *Machine Intelligence: Perspectives on the Computational Model*; Routledge: New York, NY, USA, 2012; ISBN 0815327684.

72. Chaitin, G. Epistemology as Information Theory: From Leibniz to Ω. In *Computation, Information, Cognition–The Nexus and The Liminal*; Dodig Crnkovic, G., Ed.; Cambridge Scholars Pub.: Newcastle, UK, 2007; pp. 2–17. ISBN 978-1-4438-0040-2.

73. Neil Gershenfeld Morphogenesis for the Design of Design A Talk by. Available online: https://www.edge.org/conversation/neil_gershenfeld-morphogenesis-for-the-design-of-design (accessed on 28 June 2020).

74. Sloman, A. Meta-Morphogenesis: Evolution and Development of Information-Processing Machinery. In *Alan Turing: His Work and Impact*; Cooper, S.B., van Leeuwen, J., Eds.; Elsevier: Amsterdam, The Netherlands, 2013; p. 849. ISBN 978-0-12-386980-7.

75. Nature Gene Regulation. Available online: https://www.nature.com/subjects/gene-regulation (accessed on 28 June 2020).

76. Hewitt, C. What is computation? Actor Model versus Turing's Model. In *A Computable Universe, Understanding Computation & Exploring Nature As Computation*; Zenil, H., Ed.; World Scientific Publishing Company: Singapore, 2012.

77. Rozenberg, G.; Bäck, T.; Kok, J.N. (Eds.) *Handbook of Natural Computing*; Springer: Berlin, Germany, 2012; ISBN 978-3-540-92911-6.

78. Denning, P. Computing is a natural science. *Commun. ACM* **2007**, *50*, 13–18. [CrossRef]

79. Denning, P.; Rosenbloom, P. The fourth great domain of science. *ACM Commun.* **2009**, *52*, 27–29. [CrossRef]

80. Wang, Y. On Abstract Intelligence: Toward a Unifying Theory of Natural, Artificial, Machinable, and Computational Intelligence. *Int. J. Softw. Sci. Comput. Intell.* **2009**, *1*, 1–17. [CrossRef]

81. Crutchfield, J.P.; Ditto, William, L.; Sinha, S. Introduction to Focus Issue: Intrinsic and Designed Computation: Information Processing in Dynamical Systems-Beyond the Digital Hegemony. *Chaos* **2010**, *20*, 037101. [CrossRef] [PubMed]

82. Crutchfield, J.P.; Wiesner, K. Intrinsic Quantum Computation. *Phys. Lett. A* **2008**, *374*, 375–380. [CrossRef]

83. Collier, J. Information, Causation and Computation. In *Information and Computation*; Dodig-Crnkovic, G., Burgin, M., Eds.; World Scientific: Singapore, 2011; pp. 89–105.

84. Zenil, H. *A Computable Universe. Understanding Computation & Exploring Nature as Computation*; World Scientific: Singapore, 2012; ISBN 978-9814374293.

85. Piccinini, G. Computation in Physical Systems. In *The Stanford Encyclopedia of Philosophy*; Zalta, E.N., Ed.; 2017. Available online: https://plato.stanford.edu/archives/sum2017/entries/computation-physicalsystems/ (accessed on 28 June 2020).

86. Horsman, C.; Stepney, S.; Wagner, R.C.; Kendon, V. When does a physical system compute? *Proc. R. Soc. A Math. Phys. Eng. Sci.* **2014**, *470*, 20140182. [CrossRef] [PubMed]

87. Horsman, D.; Kendon, V.; Stepney, S. The natural science of computing. *Commun. ACM* **2017**, *60*. [CrossRef]

88. Horsman, D.; Kendon, V.; Stepney, S.; Young, J.P.W. Abstraction and representation in living organisms: When does a biological system compute? In *Representation and Reality in Humans, Animals, and Machines. Studies in Applied Philosophy, Epistemology and Rational Ethics*; Springer: Cham, Switzerland, 2017; Volume 28, pp. 91–116.

89. Brooks, R.A. Intelligence without representation. *Artif. Intell.* **1991**. [CrossRef]

90. Hauser, H.; Füchslin, R.M.; Pfeifer, R. *Opinions and Outlooks on Morphological Computation*; e-book; 2014; ISBN 978-3-033-04515-6. Available online: https://www.morphologicalcomputation.org/e-book (accessed on 28 June 2020).

91. Copeland, J.; Dresner, E.; Proudfoot, D.; Shagrir, O. Time to reinspect the foundations? *Commun. ACM* **2016**, *59*, 34–36. [CrossRef]

92. Turing, A.M. The Chemical Basis of Morphogenesis. *Philos. Trans. R. Soc. London* **1952**, *237*, 37–72.

93. Kampis, G. *Self-Modifying Systems in Biology and Cognitive Science: A New Framework for Dynamics, Information, and Complexity*; Pergamon Press: Amsterdam, The Netherlands, 1991; ISBN 9780080369792.

94. Valiant, L. *Probably Approximately Correct: Nature's Algorithms for Learning and Prospering in a Complex World*; Basic Books: New York, NY, USA, 2013; ISBN 978-0465032716.

95. Simon, H.A. Rational choice and the structure of the environment. *Psychol. Rev.* **1956**, *63*, 129–138. [CrossRef]

96. Sloman, A.; Chrisley, R. Virtual machines and consciousness. *J. Conscious. Stud.* **2003**, *10*, 113–172.

97. Grossberg, G.A.; Carpenter, S. ART 2: Self-organization of stable category recognition codes for analog input patterns. *Appl. Opt.* **1987**, *26*, 4919–4930.

98. Mermin, N.D. Making better sense of quantum mechanics. *Reports Prog. Phys.* **2019**. [CrossRef] [PubMed]

99. Müller, M.P. Law without law: From observer states to physics via algorithmic information theory. *Quantum* **2020**. [CrossRef]

100. Vedral, V. Information and physics. *Information* **2012**, *3*, 219–223. [CrossRef]

101. Goyal, P. Information physics-towards a new conception of physical reality. *Information* **2012**, *3*, 567–594. [CrossRef]

102. Dodig-Crnkovic, G. Information and energy/matter. *Information* **2012**, *4*, 751. [CrossRef]

103. Fields, C. If physics is an information science, what is an observer? *Information* **2012**, *3*, 92–123. [CrossRef]

104. Wheeler, J.A. Information, physics, quantum: The search for links. In *Complexity, Entropy, and the Physics of Information*; Zurek, W., Ed.; Addison-Wesley: Redwood City, CA, USA, 1990.

105. Weizcsäcker, C.F. The Unity of Nature. In *Physical Sciences and History of Physics*; Boston Studies in the Philosophy of Science; Cohen, R.S., Wartofsky, M.W., Eds.; Springer: Dordrecht, The Netherlands, 1984; Volume 82.
106. Ehresmann, A.C. MENS, an Info-Computational Model for (Neuro-)cognitive Systems Capable of Creativity. *Entropy* **2012**, *14*, 1703–1716. [CrossRef]
107. Ghosh, S.; Aswani, K.; Singh, S.; Sahu, S.; Fujita, D.; Bandyopadhyay, A. Design and Construction of a Brain-Like Computer: A New Class of Frequency-Fractal Computing Using Wireless Communication in a Supramolecular Organic, Inorganic System. *Information* **2014**, *5*, 28–100. [CrossRef]
108. Zenil, H.; Kiani, N.A.; Zea, A.A.; Tegnér, J. Causal deconvolution by algorithmic generative models. *Nat. Mach. Intell.* **2019**. [CrossRef]
109. Cardelli, L. Artificial Biochemistry. In *Algorithmic Bioprocesses*; Condon, A., Harel, D., Kok, J.N., Salomaa, A., Winfree, E., Eds.; Springer: Heidelberg, Germany, 2009; pp. 429–462.
110. Cardelli, L.; Zavattaro, G. On the computational power of biochemistry. In *Proceedings of the Lecture Notes in Computer Science (Including Subseries Lecture Notes in Artificial Intelligence and Lecture Notes in Bioinformatics)*; Horimoto, K., Regensburger, G., Rosenkranz, M., Yoshida, H., Eds.; LNCS; Springer: Heidelberg, Germany, 2008; Volume 5147, pp. 65–80.
111. Cardelli, L. Morphisms of reaction networks that couple structure to function. *BMC Syst. Biol.* **2014**. [CrossRef] [PubMed]
112. Cardelli, L.; Tribastone, M.; Tschaikowski, M. From electric circuits to chemical networks. *Nat. Comput.* **2020**. [CrossRef]
113. Fresco, N. *Physical Computation and Cognitive Science*; Springer: Berlin, Germany, 2014; ISBN 978-3-642-41374-2.
114. Devon Hjelm, R.; Grewal, K.; Bachman, P.; Fedorov, A.; Trischler, A.; Lavoie-Marchildon, S.; Bengio, Y. Learning deep representations by mutual information estimation and maximization. In Proceedings of the 7th International Conference on Learning Representations, ICLR 2019, New Orleans, LA, USA, 6–9 May 2019.

 *philosophies*

Article

Philosophy in Reality: Scientific Discovery and Logical Recovery

Joseph E. Brenner [1,2,*] and Abir U. Igamberdiev [3]

[1] International Center for the Philosophy of Information, Xi'an Social Sciences University, Xi'an 710049, China
[2] Chemin du Collège 1, 1865 Les Diablerets, Switzerland
[3] Department of Biology, Memorial University of Newfoundland, St. John's, NL A1B 3X9, Canada; igamberdiev@mun.ca
* Correspondence: joe.brenner@bluewin.ch

Received: 31 March 2019; Accepted: 6 May 2019; Published: 14 May 2019

Abstract: Three disciplines address the codified forms and rules of human thought and reasoning: logic, available since antiquity; dialectics as a process of logical reasoning; and semiotics which focuses on the epistemological properties of the extant domain. However, both the paradigmatic-historical model of knowledge and the logical-semiotic model of thought tend to incorrectly emphasize the separation and differences between the respective domains vs. their overlap and interactions. We propose a sublation of linguistic logics of objects and static forms by a dynamic logic of real physical-mental processes designated as the Logic in Reality (LIR). In our generalized logical theory, dialectics and semiotics are recovered from reductionist interpretations and reunited in a new synthetic paradigm centered on meaning and its communication. Our theory constitutes a meta-thesis composed of elements from science, logic and philosophy. We apply the theory to gain new insights into the structure and role of semiosis, information and communication and propose the concept of 'ontolon' to define the element of reasoning as a real dynamic process. It is part of a project within natural philosophy, which will address broader aspects of the dynamics of the growth of civilizations and their potential implications for the information society.

Keywords: dialectics; epistemon; information; logic in reality; natural philosophy; ontolon; semiotics

---

1. Introduction

Philosophy, science and logic are systems of thought devised by human beings to describe their world and what it means to exist in it. In the classical West and to a certain extent in the East, throughout history, there was no separation between the disciplines. However, the value of philosophy, especially today in the West, has been diminished by several major errors: the work of Aristotle and other classical Greek, and later Western European thinkers has been misused and misunderstood, without the proper attention paid to necessary corrections and extensions made possible by modern science. The value of dialectics as the basis of reasoning, and the need for a logic based in science rather than language are major examples. In the last 100 years, phenomenology and semiotics have been proposed to bridge the gap between knowledge and reality, but all suffer from reliance on the epistemic principles of classical linguistic logic. Dialectics, in particular as expressed by Hegel, was diverted from its initial objectives and used to support limited political-economic idealism and ideologies.

The objective of this paper is to define a philosophy of and in reality that effects a 'rejunction' with some less familiar insights of Aristotle and recovers them to serve the current social objectives of the emerging information society. Rapidly, there is in Aristotle the basis not only for modern bivalent linguistic logic, but also for a logic that refers to actualizations and potentialities in a physical world of processes. We will propose an extension and development of and to the second logic of Aristotle that is grounded in modern physics. It makes possible an understanding of real processes in terms of what is essentially a *non*-Boolean logic. We will refer to two little known authors whose thought is essential to our thesis, the Russian Evald Ilyenkov and the Franco-Romanian Stéphane Lupasco. New concepts of the real dynamics of consciousness, creativity and ethics emerge from a study of these authors, making possible a rereading of the work of Western figures such as Spinoza, Kant, Peirce, Whitehead, and Heidegger and, in relation to quantum physics, Heisenberg.

We use a logic of real processes, a Logic in Reality (LIR) [1] to redefine the ontological relations between meaning, communication and language as components of a Philosophy in Reality (PIR). Under the influence of current developments in the science and philosophy of information, as noted by the Chinese thinker Wu Kun [2,3], a new more functional convergence of science and philosophy is taking place. The resulting 'dialectical realism' may make possible a more ethical development of knowledge for the common good.

1.1. Paradigms and Horizons

To describe the complex development of human knowledge, it has become customary to talk in terms of paradigms. Common examples are the scientific paradigm; the linguistic paradigm, the computational paradigm, the holistic paradigm, etc. With these are associated, more or less loosely and more or less well-defined, corresponding ontologies, e.g., digital ontology, and philosophical turns: the linguistic turn, the ontological turn and most recently the informational turn. Changes in paradigm are described as evolutionary or revolutionary.

A concept similar to paradigm is that of 'horizon'. Husserl uses the term intentional horizon to characterize the locus of the 'end' of the experience of self-awareness: "the horizon-structure of our singular empirical thought". Derrida has also called attention to its dualistic character as an opening and a limit, and Heidegger in the sense of what limits or encloses and in doing so discloses or makes available. It has also been proposed by Rafael Capurro to describe the bounds within which any existential thought or system of thought, such as our current thesis, can be made at a given time. This term has the advantage of avoiding the academic associations of paradigm, but both can be considered as applicable to our thesis, of which such dualities, and their evolution, is a key component.

Three disciplines address the forms and rules of human thought and reasoning in the generation of knowledge: logic, which has been available since antiquity; dialectics as a broader conception of logic, considered as the 'science' of correct reasoning; and semiotics which focuses on the epistemological properties of reality—the extant domain. Semiosis is the cognitive process of 'doing' semiotics.

The difficulty with both the paradigmatic-historical model of knowledge on the one hand, and the logical-semiotic model of thought on the other has been that both tend to emphasize the separation and differences between the respective domains. As discussed elsewhere below, we see this as a consequence of the retention of the basic premises of bivalent, linguistic logic separated from an adequate, physically grounded realism. Even the current holistic paradigm, as holistic science, apart from its discussion of medicine which will not concern us here, uses classical conceptions of, e.g., parts and wholes and their relation.

1.2. Forms of Thought

Human thought, its codification in knowledge and its evolution are highly non-linear processes. A key concept in our approach is that the dynamic structure of this non-linearity and in fact all non-linear phenomena have not been adequately captured by available dialectical, logical or semiotic paradigms, ancient, modern or contemporary. (A brief discussion of non-linearity is provided below). Our critique of the role of logic and dialectics per se can be summarized as follows:

- Logic has moved from being a science of nature to that of formalism for its abstract structures, exemplified in the work of George Boole [4]. The term logic is also applied in a generalized, non-scientific way to obvious regularities in the macroscopic world. Only limited relations are established to the microscopic world and its underlying quantum structure, and quantum logic remains limited to the quantum world.

- From its initial formulation by Plato-limited as Deleuze [5] has shown—dialectics has suffered from its association with Hegelian idealism, then with Marxist dialectic materialism.
- In a similar fashion as dialectics, semiotics has suffered by association with the ultimately reductionist epistemology and logic of C.S. Peirce and the lack of scientific grounding of Saussure. Noam Chomsky's generative grammar is (in our opinion) a further example of a flawed dynamic approach that still dominates linguistics.

In this paper, we will attempt to maintain a balance between the form and content of thought, in order to give proper ontological status to the former in its obvious dialectic relationship to the latter.

1.3. Objective and Outline of Paper

We are thus very much concerned in this paper with the "thinking behind the thinking". It can be seen also as a metaphilosophical contribution to knowledge, as will be clear from the reference below to the philosophy of information as a metaphilosophy. We therefore begin, in the next Section 2, with a discussion of dialectics, with the intention of showing its clear applicability to current philosophy and science. We discuss the nature and role of dialectics in knowledge and the society and recover it from reductionist interpretations. In Section 3, we present our concept of a new logic as necessary for the success of our project and an approach to a new paradigm. In particular, we propose the replacement of linguistic logics of objects and static forms by a dynamic logic of real physical-mental processes designated by Brenner as the logic of and in reality or Logic in Reality (LIR) [1]. We show that LIR has clear methodological links to dialectics and implies a metalogic that is an expression of the fundamental recursive properties of existence.

The field of information and its philosophy are having a major current impact on thought, as information involves both meaning and the physical transmission and reception of meaning. In the context of our *dialectical realism* [6], we establish in Section 4 a new set of working relations between the concepts of meaning and semiotics, a dialectics of meaning. As we will see elsewhere below, LIR provides a *naturalization* (bringing into science) of Adorno's concept. We deconstruct the concept of semiotics as an *antithesis*, an anti-dialectical system considered as one of the most general but partly misguided and misleading horizons for knowledge. We reposition prior work of Igamberdiev as a semiosis in which links to physical phenomena are stated explicitly, joining the concept of Logic in Reality as a *kind* of semiosis. We continue in this vein in Section 5, where we propose a picture of meaning in relation to the physical and cognitive phenomenon of information that is a semiosis, dialectical and logical, in our dynamic logical terms. The role of the philosophy of information as a metaphilosophy in the conception of Wu Kun is related to our new vision of semiosis.

In Section 6 we approach all four major themes—dialectics, logic, semiosis and information in terms of a new concept of the dynamic parts of existence, ontological and epistemological, for which we propose the neologisms of 'ontolon' and 'epistemon' respectively. In our Summary Conclusion Section 7, we further position our approach in relation to natural philosophy, but also make a statement of principle about the role of our project in the information society. We plan to extend these ideas to address broader aspects of the dynamics of the growth of civilizations and their potential implications for the common good.

1.4. The Scope of Philosophy in Reality

We should first state we do not presume to offer a definition of reality that would be either provable in some way or acceptable to all or most people. We refer the interested reader to the book of Colin McGinn [7] on the 'basic structures' of reality and the earlier concept of D'Espagnat of a 'veiled reality'. Perhaps the least incorrect thing one can say is that reality is something like what one thinks it is. We insist only on the existence of the dialectic between reality and appearance and the operation of the mind moving from one to the other according to the Principle of Dynamic Opposition.

To paraphrase a well-known statement in scientific theory, our philosophy of/in reality is not and is not intended to be a 'philosophy of everything'. It would be fatuous and presumptive to say that our theory covers, for example, transcendental philosophy or philosophy based on beliefs about the origin or a possible purpose of the universe. What characterizes such positions in general is their reference to an inaccessible *un*-reality, sometimes expressed by the first and last letters of the Greek alphabet, alpha and omega.

We limit the scope of our Philosophy in Reality to what is also the content of *natural* philosophy as it is currently being redefined by Schroeder and Dodig-Crnkovic [8] as well as Brenner and others. The potentially useful criterion which Brenner has proposed for delimiting natural philosophy from philosophy as a whole is its logic [9], the Logic in Reality which will be the subject of Section 3. For the prolegomenon to an integral paradigm of natural philosophy written by Igamberdiev see [10].

2. Dialectics

Dialectics appeared in Ancient Greece as a discourse between two or more people holding different points of view about a subject, but wishing to establish the truth through reasoned arguments. Formal logic—developed later—refers to subjects concerned with the most general laws of truth. This means that dialectics refers to the world consisting of multiple acting subjects, while formal logic refers to a unified sub-set of the laws common to all observers.

2.1. Dialectics in Ancient Greece

A discourse between several people providing reasoned arguments about the reality common to all of them follows from the multiplicity of the world containing interacting observers. From the logical point of Parmenides that Being is undivided and unmoved, a position which was claimed to be substantiated by the Zeno paradoxes, a search for the explanation of movement in the perceived world led to (1) substantiation of the smallest existing spatio-temporal "quanta" of movement called "atoms", referring to physical reality and (2) the formulation of cognitive pure ideal forms called *eidoi*. This separation into physical and cognitive reality came from the necessity of attributing the property of existence to Being, basically something accessible only to conscious (human) beings. This attribution is analyzed in detail in Plato's dialogue "Parmenides", which is considered as the most outstanding example of antique dialectics and in which Plato, to analyse the notion of existing Being, abandons narrowly interpreted principles of his philosophy of ideas for the goal of explaining the phenomenon of existence. The thesis substantiated in "Parmenides" is formulated in the following way: "The existing One appears as Many". Thus the existing world is present as a multiplicity that corresponds to the principle of multiplicity of primary elements called "seeds" in the philosophy of Anaxagoras. These were later defined by Aristotle as *"homoeomeria"*. They include both potential and actual constituents and are harmonized in the *omnium* of their relations by the principle called *"Nous"* by Anaxagoras. A similar philosophical concept was formulated in modern times by Leibniz, who called these elements "monads". Below, we will discuss monad in relation to our definition of an "ontolon" as the element of real existence. The definition of the principle uniting *homoeomeria* as *"Nous"* assumes that a non-physical 'discourse' reasoning between them takes place that results in their harmonization in the Universe. Such harmonization can be either based on the ideal pre-established principles defined by Leibniz as a "pre-established harmony", seen in antiquity in the concept of the Demiurges of Plato's dialogue "Timaeus", or it can be achieved through a kind of a principle of natural selection that was formulated explicitly in the poem *"De rerum natura"* of the Roman poet Lucretius, who followed the atomistic philosophy of Epicurus.

The philosophy of atomism of Leucippus and Democritus initially appeared as a response to Parmenides and Zeno as an attempt to introduce movement in the world. At first glance, it avoided dialectics by claiming atoms as real existing elements, between which was emptiness (the vacuum). In the early versions of atomism, atoms were not considered as ideal essences that could be involved in a kind of discourse, although there is some evidence that Democritus sometimes called atoms "ideas". The strict determinism of the Democritean universe does not leave room for the possibility of a 'discourse'. Later, Epicurus postulated an unpredictable movement (swerve; swerving) of atoms called by Lucretius *"clinamen"* which introduces potentiality in the universe of atoms and makes it more diverse and variable. These deviations initially appear arbitrary, but they can be controlled resulting in higher levels of organization. In this picture, finally, human consciousness appears which controls them further. The atoms of Epicurus can be viewed as quanta in the process of actualization and the actualized Being is Multiplicity due to the *clinamen*. Without *clinamina* that enlarge the field of potentialities, a strictly deterministic universe would not be able to evolve and generate a multiplicity of events, phenomena and realizations. However, the principle defined by the term *clinamen* cannot be well developed without the more advanced concept of potentiality and actuality introduced in Greek philosophy, by Aristotle despite the fact that Epicurus lived after Aristotle. Aristotle developed his concept of potentiality without reference to atomism, and Epicurus was not influenced by Aristotle's understanding of potentiality.

The most important point here is that atoms possessing *clinamina* escape from the strict determinism of the Democritean world. They can enter into a process that reduces the field of potentialities defined by their *clinamen* and form a kind of discourse or 'communication'. This discourse is not strictly physical as it escapes physical determinism being based on the consistency during interaction. The result of such interaction is a *consistent history* that is produced in it[1]. The complexification of interaction is associated with generation of predictability of the movement defined as *clinamen* and the possibility of its control. This corresponds to the development of cognition in systems of interaction of atoms possessing *clinamina*.

By the introduction of *clinamina* as the properties of atoms, it becomes apparent that the atoms interact not only via application of deterministic forces but also via the influence of the potential fields of their possible unpredictable deviations, recursively modifying their capacity for such unpredictable movement. Moreover, if atoms unite in organized structures, these capacities become restricted to a certain extent. The interaction via the effects of *clinamina* on each other can be described as a mutual measurement process, and when atoms become united into organized structures this measurement appears as an internal measurement within the structure. This kind of measurement can lead to further complexification of the system via reduction of the potential field and self-referential 'memorization' of such reduction.

The significance of Aristotle in understanding such a discourse involving potential reality makes him the greatest figure in dialectics. The common opinion is that Aristotle mainly developed formal logic which is the content of his Categories (part of his *Organon*), but he also introduced a kind of logic in reality (see below the logic of/in reality of Lupasco and Brenner). This includes the transition from the potential to the actual, although the opposite transition is not analysed in detail. The great achievement of Aristotle is the inclusion of *potentia* in his second logic (dialectics). This is seen in his *Metaphysics* and in more detail in *De Anima* (On the Soul). Heredity can be understood in Aristotelian terms as the transfer of information, defined as "entelechy as knowledge" in a seed as compared to the realized "entelechy as the actual exercise of knowledge" of the developed organism (cf. G. Stent's *Molecular Genetics* [11]. Matter in Aristotle's conception is a pure potentiality which is a prerequisite of two types of actuality: one is information ("entelechy as possession of knowledge") and the other is actual realization/presence ("entelechy as the actual exercise of knowledge"): "Now the word actuality has two senses corresponding respectively to the possession of knowledge and the actual exercise of knowledge" [12]. In the latter ("actual exercise"), a selection from two or more possible realizations takes place according to the first type ("possession of knowledge"), and the discourse becomes incorporated in the total reality instead of being a result of interaction of reasoned arguments of conscious subjects. Being in the philosophy of Aristotle includes the unity of potential reality and its realizations, a concept which was developed by him in detail but which had arisen in the concept of

---

[1] Attempts have been made to make a 'consistent histories' approach to quantum phenomena. We consider these as tautological and ultimately reductionist. As we will see in Section 3, our approach assigns ontological value to *inconsistency*.

the primary substance as *apeiron* outlined more than two centuries earlier by Anaximander. According to Carlo Rovelli [13], Anaximander was the founder of scientific thought in human civilization.

Another important aspect is the understanding of a soul as a *capacity*, inseparable from the body, which we later find in the concept of substance in the philosophy of Spinoza. According to Aristotle, " … the soul neither exists without a body nor is a body of some sort. For it is not a body, but it belongs to a body, and for this reason is present in a body, and in a body of such-and-such a sort" [12]. On the other hand, different bodies can be animated by the same set of capacities, by the same kind of soul, so soul can be used in a singular, not plural, sense. This was earlier substantiated in Plato's dialogue "Parmenides" formally, while Aristotle presents this as a foundation of a kind of natural science which was later defined as psychology. The discourse of different potential realizations becomes incorporated into the reality that evolves from the inanimate potential matter to the actual realization, bearing form and being capable of an information transfer through the discourse of the substance with itself and in which it appears as a multiplicity of "seeds" sharing the same "soul".

2.2. Dialectics in Modern Times

In modern times, the logic of discourse was incorporated into philosophical thought following the main feature of the new European philosophy that was formulated by Rene Descartes as the distinction between a *res cogitans* and a *res extensa*. Being initially represented as two separate essences, they were unified in different philosophical systems in different ways, and dialectics appeared as the discourse for establishment of such unification. Thus, we can see the development of dialectical principles in the systems of Spinoza, Leibniz and Kant. After Kant, dialectics became associated with Hegelian idealism on the one side and Marxian materialism on the other. In this Section, we will discuss these advances in modern times, their limitations and possible future developments.

Another important aspect here is that if substance is a *causa sui*, it cannot be fully cognizable, since only mechanisms can be fully cognizable in the framework of mechanics. That is why Spinoza ascribed to substance an infinite number of attributes, among which only cognition and spatio-temporality are perceivable by us. In relational terms, only these two attributes can be involved in relations, while other attributes cannot be cognized although they can be involved in some sense in the shaping of the *res potentia*. The statement of the infinite number of attributes by Spinoza has similar meaning to the statement of Leibniz that "monads have no windows" (there is no window through which to see these attributes) and to the statement of Kant about the non-cognizability of the *Ding-an-sich*. The latter also appears as a monad and should not be mixed with the physical object, as it cannot be reduced to its spatio-temporal structure or to its ideal form. Later in this paper, we will call this primary unit "ontolon". The "non-cognizable" attributes shape the potential of monad/Ding-an-sich/ontolon. These attributes correspond to the Epicurean *clinamen* which, as noted above, is beyond shape and form.

The great progress in philosophical thought made by Immanuel Kant, which he called the Copernican revolution in philosophy, was the revelation [14] that mind is not the basis of Being but only the instrument of human cognition. Real Being is beyond mind and can be defined as a *Ding-an-sich* (thing-in-itself). The later development of German philosophy (Fichte, Schelling and Hegel) abandoned this basic statement and returned to a Mind in its totality that generates the material world as an *Anderssein (Andersheit)* in its dialectical discourse. On the contrary, Arthur Schopenhauer associated the *Ding-an-sich* with the primary energy called *Wille* (will) whose permanent goal is actualization *via* generation of representation (*Vorstellung*). Another trend, which overturned Hegelian idealism, resulted in Marxism in which dialectical discourse is located in an inanimate reality called matter. However, Marxism did not develop the concept of substance in the sense of Spinoza or other earlier philosophers. Dialectical discourse appeared, in particular, in Engels' interpretation, as a vaguely formulated set of "laws of dialectics", which emerge without proper substantiation.

Friedrich Engels, in his unfinished work *Dialectics of Nature*, formulated "three laws of dialectics" [15]. He elucidated these laws as the immanent properties of dynamics of material substance, although his concept is, however, not elaborated well as compared, e.g., to Spinoza. These "laws" are the following: (1) the law of the unity and conflict of opposites (which arises in Heraclitus); (2) the law of the passage of quantitative changes into qualitative changes (generally based on the ancient paradox of the heap), and (3) the law of the negation of the negation (which may be considered as the invention of Hegel). In fact, in comparison to Hegel's philosophy where dialectics appears as an internal discourse of the Absolute Idea, in Engels interpretation dialectics is positioned far away from any discourse and represents a set of vaguely and rather reductively defined formal laws. This became the basis of "dialectical materialism".

To realize discourse, an element of the world should represent a monad that performs some kind of internal 'computation', which in the more materialistic view can be defined as an "ontolon" (see below the discussion on ontolons in Section 6). In this sense it is more logical to discuss "dialectical organicism"[2] as suggested by Joseph Needham [16] than "dialectical materialism". In the organicism interpretation, the dialectical discourse of simple monads, ontolons in our terms, generates the complex structure of space-time, and, as in Spinoza's philosophy, this discourse rises to the *causa sui* principle. The formulation of "dialectical materialism" without clear definition of the concept of substance as matter that would justify the *deductive necessity* of the "laws of dialectics" resulted in the difficulties of development of this concept by the next generations of philosophers. We will show later that the concept of an ontolon can resolve the difficulties in the ontological interpretation of dialectics in nature.

We will turn now to the "post-Marxist" philosophers Merab Mamardashvili [17] and Evald Ilyenkov [18] who performed the most important re-evaluation and development of the Marxian dialectical concept of consciousness. Since the developed concept of substance is lacking in Marx' theory, Merab Mamardashvili refers to Descartes and Kant in its description, while Ilyenkov is grounded mostly in Spinoza. Both Mamardashvili and Ilyenkov refer rather to the logic of *Das Kapital* of Marx than to the *Dialectics of Nature* of Engels. To what extent the newly formulated principles really arise in Marx or are the result of the major reformulation performed by Mamardashvili and Ilyenkov is not so important, but in our view their interpretation goes far beyond the basic formulation of Marx. We will outline first the approach to dialectics as it was formulated by Mamardashvili and then turn to Ilyenkov.

According to Mamardashvili (1930–1990), who significantly contributed to the rationalist theory of perception whose origin was in Descartes and Kant [17], the relation of subjective signifying consciousness to objective reality (the set of signified material bodies) is mediated by the potential set of the whole system of, in particular social and political relations organized hierarchically. This means that Marx, according to Mamardashvili, discovered the phenomenological nature of consciousness via its quasi-objective nature by introducing an abstraction that allows the analysis consciousness as the objective transformation of external objects into quasi-objective patterns, without direct involvement of the processes taking place in internal subjective reality. This means that the nature of consciousness is placed beyond the phenomena that serve for the maintenance of the social system of communication. "Being–consciousness" becomes unified, so that Being and consciousness appear as the different aspects of one continuum in which the object and the subject, the reality and its representation, the real and the imaginary are not strictly separated, while remaining relatively differentiated and non-identical to one another. They are connected in the continuum via the relational operator of transformation realized in the course of social dynamics. This is represented as the "dialectical" nature of consciousness in which the discourse between conscious subjects, mediated by the incorporation of the actualized material reality into this discourse, generates the conditions for social dynamics and progressive social evolution. The actualized material reality appears primarily as the result of previous human activity that has formed the signified social memory that represents the basis on which the current social structure is built. This "basis" is not only material but also cultural, and it shapes social structure by providing

---

2   The application of the principles of Logic in Reality allows one to cut through the endless discussion of organicism *vs.* realism *vs.* holism *vs.* reductionism. Complex dynamic part-whole relationships are possible, as in the first concept, without the system being 'alive', but which are not reducible to their energetic substrates.

already existing forms or models. In other words, it "geometrizes" the society in the same way as Spinoza's substance geometrizes the world in the course of its self-actualization.

2.3. Dialectics in Ilyenkov's Conception and Beyond

The concept of dialectics as a metalogic based on reasoning and discourse was developed by Evald Ilyenkov (1924–1979), who incorporated its principles into basic substance understood in the sense of Spinoza and having the basic property of a *causa sui*. Despite his close association with Marxist dialectic materialism and apparent rethinking of Hegelian idealism, Ilyenkov suffered from the attacks of both orthodox Marxists and anti-Marxists, which resulted in his premature death. According to Ilyenkov, substance perpetually generates objective forms of subjective activity which follow a logic external to a material body. In this regard, the process of cognition which includes discourse and reasoning is not transcendent to the being, but immanent to it. Before Ilyenkov, such a point of view was formulated in the Marxist psychology of Lev Vygotsky and Leontiev in the 1930s, later substituted by a reflexology grounded in the works of Ivan Pavlov. The main idea on which Ilyenkov's concept of dialectics is based is the unity of cognition and space-time, of a *res cogitans* and a *res extensa*, which are linked via a *res potentia* (see below). The unity of *res cogitans* and *res extensa*, which is the central point of Spinoza's philosophy, is attributed to substance from the lowest levels of its organization and becomes highly expressed at the highest levels such as human civilization. The basic property of cognition, following Spinoza, is a capacity of a body to build a trajectory of its movement across other bodies according to a logic of arrangement of these bodies in the space external to that body. The idea of a thing in this regard becomes fully coincident with the way of its being, which is its generation based on this idea. In other words, the idea as *eidos* turns to be the idea as *technos* as defined by Mamardashvili [17], and these two aspects of idea (eidos/technos) are inseparable in the generating activity. As an example, a geometric shape is ideal because it represents a way of formation/generation of all material bodies possessing this shape.

The basic function of intellect, according to Spinoza and Ilyenkov, is to move and arrange external objects. Humans perceive the world only because they move and arrange external bodies in their activity. This activity is not based on the mechanic causality but on the causality which immanent to the primary substance which is *causa sui*. Mechanical causality is always external to the body, while substance possesses causality in itself realizing self-movement via establishing relation to external bodies by its abstracting capacity which generates pure (ideal) forms of reality. In this activity, the *ordo et connexio idearum* coincides with the *ordo et connexio rerum*, in other words, ideal activity results in valid practical implementations. At the level of social organization, intellect is involved in establishing relations not only to external material objects but to other intellectual beings, which becomes the basis of morality and successful social communication.

The unity of cognition and space-time needs to be incorporated not only in philosophical thought but also into the foundations of mathematics. This has been realized only recently in the concept of meta-mathematics developed by Voevodsky [19] who included geometrical foundations of mathematics in its basis. The intrinsic logic of meta-mathematics corresponds in this approach to the spatio-temporal structure that is generated internally on the basis of this logic. When geometry is introduced into the foundations of mathematics, the world becomes shaped in a particular way fitting its habitability, which resembles the anthropic principle in physics. The limits of geometry become associated with the limits of computation of the particular world, and in the theory of homotopic types developed by Voevodsky the basic foundations of mathematics can be verified computationally [19]. The grounding of Ilyenkov's dialectics in Spinoza's concept of substance represents its major advance, but it has certain limitations due to the apparent disregard of relational principles in the operation of cognitive activity. In fact, Spinoza's substance is a manifestation of the 'One', while Plato, in the dialogue "Parmenides", had already substantiated that the existing One appears as Many. In the universe of the forms of existence or "ontolons", the spatio-temporal order appears as a relation between objects established in their interaction. The intrinsic limits of computation shape the spatio-temporal order and also pose limitations on cognitive activity. They represent the principles that are inherent to our world and may result from the transcendent action of the establishment of the actualized physical world from the pure logical principles that are insufficient for its appearance. To what extent this transcendent action is similar to our cognitive activity remains open, but in this paper we present what amounts to an 'immanent' alternative in the work of Lupasco. David Hume [20] indicated a possibility of such similarity. If the basic property of consciousness, according to Spinoza and Ilyenkov, is the establishment of space-time, then the introduction of fundamental constants at the birth of the Universe is a 'conscious' act that sets the limits of (and for) actualization of the *res extensa* from the *res potentia*. The assumption of a *res cogitans* here seems apparent, otherwise we need to introduce the principle of natural selection between universes, as in the multiverse of Smolin [21] which assumes the unsubstantiated actual pre-existence of them all. The anthropomorphic 'bootstrapping' by a basic 'intelligence' that determined the limiting conditions of the physical world is, however, beyond scientific reasoning and cannot be discussed in this paper.

3. Logic in Reality (LIR)

On many occasions, Brenner has discussed the principles of Logic in Reality and their derivation from the logical system proposed by Lupasco, for example in articles on consciousness [22], ecology [23] and natural philosophy [9]. Underlying this work is a vision of Lupasco's Principle of Dynamic Opposition (PDO) as operative in nature. We reproduce from [1] the best and simplest expression of this principle: "The antagonistic dualities of our world can be formalized as a structural, logical, and metaphysical principle of opposition or contradiction instantiated in complex higher-level phenomena (Principle of Dynamic Opposition   PDO). The fundamental postulate of LIR is that for all energetic phenomena (all phenomena) alternate between degrees of actualization and of potentialization of themselves and their opposites or 'contradictions' but without either going to the absolute limits of 0% or 100%. The point traversed at which a logical element and its opposite are equally actualized and potentialized is one of maximum interaction from which new entities can emerge. It is designated by Lupasco and Basarab Nicolescu, the physicist colleague and major continuator of Lupasco [24], as a 'T'-state, T for included middle or third (*Tiers-inclus*). A relatively simple example of a physical T-state is the transition state in a chemical reaction. This is the point at which the number of molecules of reactants moving toward more thermodynamically favored products and the number moving in the reverse direction is the same. We use the concept of T-states to evaluate both philosophical and scientific theories, including patterns of human individual and social behavior. A dynamic systems view can be used to focus on the feedback or recursion present in all natural processes."

The absence of debate of these concepts of the dynamic general properties of natural phenomena has given them the appearance of a statism when the opposite was intended. The choice of the domains in this paper—dialectics, semiosis and meaning/information, to which the PDO is intended to apply, is in part based on the difficulties of explaining them through the use of standard doctrines, as well as the desire to 'mobilize' the PDO as an explanatory methodology. As Brenner suggested at the 2015 Information Summit Conference of the International Society for Studies of Information in Vienna [25], the concept of scientific method is only one of those meaningful in the contemporary practice of science. Computational methods can be and are applied routinely in all the sciences, but their limitations demand a directed interpretation. In the human domain, it is the application of operative or organizational principles to an individual or social cognitive process to determine its dynamics, what "forces are at work", that we consider essential for the determination of an informational commons.

In a conception discussed at the Vienna Conference, the Information Society is at three crossroads in terms of its future development: we quote from Brenner's paper [26]: (1) a Socio-Political Crossroads where trends toward improvement in the quality of life are offset by a regression and degradation of the mental and social environment, both in part due to the massive role of information in the society; (2) a Transdisciplinary Crossroads, where the science and philosophy of information as disciplines may develop in the direction of integration in an Informational Turn, a new way of Informational Thinking as proposed by Wu Kun [27] that can support efforts toward a Global Sustainable Information Society, in the term of Wolfgang Hofkirchner [28]. Alternatively, they may diverge or regress in the direction of increasingly socially irresponsible specialization and scholasticism; and (3) a Metaphysical Crossroads, inseparable from the first two, involving the direction of development of the science and philosophy of information as a metaphysics, a crossroads that includes a definition of the dynamic relation of man to the universe. Like the other two, there is a positive branch leading toward less dysfunction at the individual and social level. The negative branch implies an on-going blockage of ethical development of the society.

In our view these three domains are also at their own crossroads: they can continue in separation or accept a logic and methodology that places the emphasis on their non-separability and co-evolution. The dialectic logic of/in reality outlined above offers a methodology that provides for linking them dynamically.

3.1. The Philosophical Logic of Stéphane Lupasco

In the broadest possible sense, this is a paper about change, better about changing and processes in the real world. We are interested primarily in ontological change, about which little has been written from a logical standpoint. The reason is obvious: change is ubiquitous in existence and experience. Theories of change, however, have focused on making it mathematically, computationally and logically tractable, within the framework of standard logic.

Differential equations provide an excellent description of continuous change, but what if the change in question is partly discontinuous, recursive and/or random? In fact, change is contradictory: the most familiar thing about change is that it never occurs in isolation from stability. Change is regular *and* irregular; consistent *and* inconsistent; continuous *and* discontinuous. Since the only logics available have been propositional bivalent logics, incapable of accepting real contradictions, they have been incapable of describing change.

Brenner has written several articles in the last decade which describe the non-standard, non-linguistic logic that we see instantiated in changing processes [29]. It is based on the logical system proposed by the Franco-Romanian thinker Stéphane Lupasco (Bucharest, 1900–Paris, 1988). The extension of Lupasco's system made by Brenner is called Logic in Reality (LIR), and it has been applied most recently to the fields of information [30] and the philosophies of information and ecology, as noted above.

Despite its publication in some fifteen books in his lifetime and its continuation by his associate Basarab Nicolescu [24], Lupasco's system of thought has remained unknown outside France, where it had been rejected by the academic community. With a few notable exceptions, Brenner's publications have suffered the same fate. The reason here is less obvious, but in our opinion it has to do with the fact that acceptance of our logic of and in reality requires the acceptance of a new, scientifically grounded concept of the dynamics of change in all complex, interactive phenomena at biological, cognitive and social levels of reality.

The comparison of the Lupasco logic with standard logics is rendered difficult due to the limitation of the latter to the linguistic domain. These semantic logics, bivalent or multivalent and their most recent epistemic, paraconsistent and paracomplete versions still require absolute separation of, for example, continuity and discontinuity, space-time and matter, chance and necessity, etc. and lead in linguistics to the paradoxes with which we are all familiar. Paraconsistent logics, which accept contradiction, capture only the linguistic as opposed to the physical aspects of processes, although some real inconsistencies in simple change (Sorites problems) are accepted. Logics of epistemic change are based on linguistic abstractions. No logical characteristics are ascribed to the physical processes of change. It is thus not surprising that no generally applicable theory of change has been developed for the extant domain of macroscopic, complex and interactive processes.

3.2. Logic in Reality: Axioms and Categories

As noted in Section 2, Aristotle outlines his fundamental concept of potentiality and actuality in the *"De Anima"*, (*"On the Soul"*). In this already metaphysical context, he writes that potentiality is matter and actuality is form. Thus, we need to know what matter and form are, but we can assume they are different. Why, to use a more modern term, potentiality should be instantiated in matter and actuality in form is not clear. What is more important for this study, however, is the relation that *Aristotle* sees between potentiality and actuality. In *De Anima* there is no direct indication that one can be transformed or transform itself into the other, but this is implied by his concept of life as an internal transformation of the body. If it is implied by Aristotle's view of the movement of energy, there is certainly no indication of the possibility of movement from actuality to potentiality.

As discussed by Brenner in [1], Lupasco's definition of the two terms in terms of energy has place for this contrary forward movement. It is implied by the fact, observable in part, that no complex process goes to the absolute limits of 100% potentiality or 100% actuality, except in trivial cases or those in which there is no real interaction. The entire literature around Schrödinger cat fails as 'alive' and 'dead' are not interacting states. Given this interpretation, why is not more attention paid to it in philosophy when it fits modern physics?

3.3. Toward a New Non-Boolean Logic

Since George Boole published his *Laws of Thought* [4] in 1854, Boolean logic has been the canonical logic of science and philosophy. Boole demonstrated that classical, bivalent propositional logic could receive an algebraic formulation and proposed a general symbolic method for logical inference. His algebra contains terms for both quality and quantity and provides the basis for standard probability theory. However, his terms for *quality* are strictly limited to formalizable, binary properties of phenomena. Non-Boolean logics and algebras [31] have been shown, relatively recently, to be really necessary in the area of quantum mechanics, with a few interesting but constructed exceptions in the work of Diederik Aerts [32], Elio Conte [33] and others.

However, as should be more widely appreciated, Boole himself was aware of the limitations of his own system and was completely open to others, sometimes of a striking modernity: "we sometimes find more just conceptions of the unity, the vital connexion and the subordination to a moral purpose, of the different parts of truth, among those who acknowledge nothing higher than *the changing aspect of collective humanity* (italics ours), than among those who profess an intellectual allegiance to the Father of Lights". Further in a key Appendix to [4] he writes that the central role of mathematics as derived from his Laws of Thought, it is not a sufficient basis either of knowledge or of discipline, "As truly, therefore, as the cultivation of the mathematical or deductive faculty is a part of intellectual discipline, so truly it is only a part."

Balance is necessary in any view of the operation of the human mind: "I would especially direct attention to that view of the constitution of the intellect which represents it as subject to laws determinate in their character, but not operating by the power of necessity; which exhibit it as redeemed from the dominion of fate, without being abandoned to the lawlessness of chance," Boole's logical laws of thought can manifest their presence "otherwise than by merely prescribing the conditions of formal inference." The distinctions between true and false, between correct and incorrect, are cornerstones for his and all other standard logics. But this distinction "exists in the processes of the intellect, not in the region of physical necessity".

Boole is honest in admitting not to have found a *construens* to accompany his *destruens*, but several of his remarks suggest that some aspects of Logic in Reality would have been congenial to him. One was to the effect that his Laws of Thought (logic) were capable of precise scientific expression, but were invested with a lower 'authority' than the laws of nature in general. "Were the correspondence between the forms of thought and the actual constitution of Nature proved to exist, whatsoever connexion or relation it might be supposed to establish between the two systems, it would in no degree affect the question of their mutual independence." "Wherever the phenomena of life are manifested, the dominion of rigid law in some degree yields to that mysterious principle of activity", a teleology accomplished "not, apparently, by the fateful power of external circumstances, but by the calling forth of an energy from within." We quote Boole *in extenso* [4], because we have seen no references to these passages elsewhere. It has been left to Lupasco to continue and talk about the 'laws of energy' and their deployment and to show that some real macroscopic processes follow a logic whose terms do not commute or distribute and are accordingly non-Boolean. This extension of logic, as noted, has not been widely accepted in the literature.

3.4. The Link to Dialectics

As stated by Lupasco, dialectics can be considered neither more, nor less, than the generalization and mental expression of conflicts in nature and civilization, and their resolution, that man has observed from time immemorial. "Beings and things seem to exist and are able to exist only in function of their successive and contradictory conflicts" [34]. For Heraclitus, conflict did not mean the splitting or destruction of the unity of reality, but its constitution. The *logos* is the only "abiding thing", the orderly principle according to which all change takes place as a 'binding-together'. Conflict (*polemos*) and *logos* are the same. As defined in Lupasco's Table of Deductions, ortho- and para-deductive chains of implication are thus an integral part of logic. A *disjunctive* dialectical oscillation is required between the first three implications of implications, and between the three implications of implications of implications controlled by the former, and so on, of which the following is the first sequence (A = actual; P = potential; T = T-state):

$$[(\supset A) \supset (\overline{\supset} P)\,] \lor [(\overline{\supset} A) \supset (\supset P)\,] \lor [(\supset T) \supset (\overline{\supset} T)\,]$$

This implication formula defines the meaning of disjunction as the mechanics of dialectics: no dialectic without disjunction and vice versa. It is disjunction that is implied by the fundamental postulate that permits the dialectic, and the dialectic implied by the same postulate, as principle of antagonism that permits and requires the disjunction, the connective 'or' [35]. In an early paper [36], Aerts describes the failure of the classical 'or' but offers no adequate replacement.

The operation of the fundamental structural logical principle of LIR implies a type of *dialectics* at all levels of reality between the two terms of any interactive duality. In other words, the dialectical characteristics of energy—actual and potential, continuous and discontinuous; entropic and negentropic, identifying or homogenizing and diversifying or heterogenizing—describe the dynamic structure of all interactive phenomena, physical and mental, including information, propositions and judgments.

We quote here another key concept in Lupasco [34]:

"Energy must possess a logic that is not a classic logic nor any other kind based on a principle of pure non-contradiction, since energy implies a contradictory duality in its own nature, structure and function. The contradictory logic of energy is a real logic, that is, a science of logical facts and operations, and not a psychology, phenomenology or epistemology."

Contradictions or dynamic oppositions thus exist in things being continuous and discontinuous, unified and diversified, wave and particle, local and global, in some way at the same time, but not completely so, only in the sense of alternating between ideal limits which are never reached. A standard Aristotelian logic—one which tries to eliminate or avoid contradiction of any kind—is not adequate to describe real systems, all of which are derived from energy in all its *macroscopic thermodynamic* forms: mechanical, electromagnetic, electrostatic (chemical). LIR is not intended to apply to gravity and quantum entities as such. Some paraconsistent logics permit true contradictions, but retain idealized, abstract concepts of truth and falsity. Consequently, they fail to give an adequate picture of the emergence of complex, real-world phenomena. These points apply to all phenomena: ideas, theories, propositions, as well as physical systems.

Contradictions, in the physical sense of real opposing processes, entities or properties can never disappear completely, since this would imply, ultimately, going below the quantum limit, defined by the Planck quantum of action. All phenomena thus continually but non-reflexively (that is, without 'perfect' circularity) alternate between degrees of actualization and of potentialization of themselves and their opposites or contradictions. The application of the logic of the included middle implies an open, incomplete structure of the set of all possible levels of reality, similar to that defined by Gödel for formal systems[3]. Concatenations of systems and dialectics in the sense of a Hegelian or Marxist synthesis never yield a third term. Hegel's vision was solely philosophical, without the advantage of the physical grounding we now have. The Lupascian T-state is not a static term, but a dynamic state, and emergent T-states, at a higher level of reality, can also enter as elements into contradictory relations. In a sense, which remains to be explored in greater detail, *relations are T-states*, conceptualizations of change that are nonetheless energetic within the laws of physics. This scheme reflects only one step in what Lupasco called the 'ortho-dialectic' processes of processes that constitute change, looking from the process standpoint. From the point of view of the entity, since no real process returns to the same point, if the process is going in the direction of non-contradiction (of diversity or identity, the net result is that it will be more of an identity or more of a diversity in consequence. In this scheme, the process that leads to more and more differentiated individuals, that is, biological processes, is one of heterogenization which should be distinguished from the contradictory process that creates homogeneous individuals from a multiplicity of entities.

The fundamental axioms of LIR imply that entities can be *both* the same and different, *both* distinguishable and indistinguishable. This seems consistent with the interpretation of Krause [37] for quantum cases. More formal ways are still needed, however, to distinguish between macroscopic process elements involved in an 'active' process and objects for which the dialectics are 'frozen': subject to an input of energy, they are to all intents and purposes in the 'classical' part of the LIR theory. This is similar to the quasi-set situation, for macro elements that are distinguishable; the set-theoretical description has a classical part[4].

Jainist philosophers in India [38], in the first half of the 1st Millennium made similar statements in a positive mode: quoting from Stcherbatsky, the nature of reality, they said, is permanent and impermanent at the same time, finite and infinite, particular and universal. They realized that a being with absolute identity would be unrelated to all others and could not exist, but without some identity, it would be indistinguishable from everything else. Many authors use this construction when they are unable to provide a satisfactory explanation of the phenomenon under study. However, no explanation is given of how such states of affairs might be instantiated, and the phrase 'both at once' can only be understood metaphorically.

The performance of philosophy can thus be considered as a dynamic and dialectical process itself, in which one oscillates between analytic and synthetic approaches, each serving as a control of the other. In the LIR conception, all physical processes, including mental or neuro-psychic, are first of all real *qua* the energy involved in their instantiation. To think about something being an illusion or an abstract entity requires energy to do so. The logic of/in reality proposes a dialectical relation between 'reality' and its appearance to a conscious observer. It is the totality of this picture that we consider realism; reality and appearance are both real. What is *not* real then is not in the sense of lacking any character of dynamic opposition, that is, non-spatio-temporal phenomena such as abstract entities of all kinds. We see here the basis for our 'discovery', outlined below, of a criterion for distinguishing between natural philosophy and all philosophy.

A standard anti-realistic argument is that since perception can be and often is unreliable, realist theories such as LIR that that are based on it are invalid. We consider treat experimental discovery, as for example the components of perception, as *generally* empirically reliable, and, subject to consensus,

---

[3]   Computational logic now includes concepts of formal systems as open, capable of handling changing or evolving information, replacing the Hilbert concept of formal systems as closed.

[4]   This is again similar to the contextual concepts of Aerts. It should be considered the rule rather than the exception that macroscopic systems as well as quantum systems have classical and non-classical parts.

an adequate naturalistic philosophical explanation of why our beliefs based on perceptions represent knowledge about objects that are independent of those perceptions to all intents and purposes. The availability of what Brenner has called catastrophic counterexamples will not change this conclusion. Accordingly, any change to a new theory can preserve structural properties allowing a certain ontological continuity accompanying a conceptual revolution as discussed by Cao [39]. This ontological *synthesis* is a dialectical picture of growth and progress in science that reconciles essential continuity with discontinuous appearance in the history of science, a process that, again, is a logical one in LIR.

3.5. Logic in Reality and Hegel

The major precursors of the logic of/in reality are discussed in [1] and in several papers by Brenner. However, because parallels to Hegel's dialectics, logic and ontology exist, we will state explicitly how LIR should be differentiated from Hegel's system. Lupasco considered that his system included and extended that of Hegel. However, one cannot consider Lupasco a Hegelian or neo-Hegelian without specifying the fundamental difference between Hegel's idealism and Lupasco's realism. Both Hegel and Lupasco started from a vision of the contradictorial or antagonistic nature of reality; developed elaborate logical systems that dealt with contradiction and went far beyond formal propositional logic; and applied these notions to the individual and society, consciousness, art, history, ethics, and politics.

However, Lupasco proposed *two* dialectics, ascending and descending (*diverging*) toward the non-contradictions of identity and diversity and a *third* dialectics *converging* toward contradiction. As above, the ubiquitous contradiction in nature is that inherent in energy and is the only existent reality. To say that material-energetic reality was the result or emanation of some other necessity as the foundation of the real amounts to tautology or mysticism, and Hegel's "obscure logical descriptions remained without a future for logic and science". As Lupasco expressed it, Hegel's system was "only half of a dialectics" [40]. In Hegel, the affirmative value of identification always transcends the negative value of diversification. In LIR, contradiction between them, and what can emerge from them, is established at the basic physical level.

As pointed out by Taylor [41], Hegel's thesis depends on a premise of ontological necessity that in turn would depend on a contradiction of the finite. Hegel established or expounded his resulting ontological structure at 'high' levels, but his project required demonstration of his ontology at the lowest level of simply determinate beings, and his attempted proof of contradiction failed. The realism of LIR successfully answers this major objection to the coherence of Hegel's system, without requiring a commitment to the idealist part of his doctrine.

The Hegelian philosophical vision of "embodied subjectivity, of thought and freedom emerging from the stream of life, finding expression in the forms of social existence, and discovering themselves in relation to nature and history" is still relevant. However, Lupasco's view of contradiction founded a dynamics, whereas Hegel's did not, precisely because his system is *not* metaphysically and physically grounded at the "lowest level of simply determinate beings" that is, microphysical entities. Lupasco [34] showed that there is no *deductive* necessity in Hegel for thesis generating anti-thesis, let alone any subsequent fusion[5]. In line with our effort of naturalization of philosophy in general, LIR can be considered as Hegel *naturalized*, since a physical basis in reality for Hegelian change has been defined.

We note with regret the absence of any resonance of Lupasco's work in French thought, other than in the relatively non-scientific context of the International Center for Transdisciplinary Research founded by Nicolescu, Lupasco and a few others. Lupasco lost his position in the French National

---

[5] Lupasco rejected Hegel's dynamic relation between being and becoming, since he wanted to limit contradiction to the domain of becoming, which drastically limits the value of Lupasco's thesis. In fact, Lupasco's universe consists of almost nothing but Becoming as functional contradiction, the alternation of the actualization of a phenomenon, with the potentialization of its contradiction, and the actualization of the former, plus emergent T-states. Contradiction is absent only in affect or affectivity, which has no energetic aspects and is the only constituent of being. This metaphysical position is incompatible with the non-naïve realism of LIR.

Scientific Research Center (CNRS) because no one could decide in which field his logic and approaches to biology and psychology should be placed. In the 1980's his work was referred to disparagingly as just 19th Century German romanticism. In a competition for a key position in the Collège de France in the 1950s, Merleau-Ponty was chosen over his contemporary rival—Lupasco. The marginalization of Lupasco can be dated to this event. More recently, no mention of Lupasco is to be found in the 'dialectical walk in the sciences' of Évariste Sanchez-Palencia [42]. He wrote in 2012 that if dialectics is defined as the "general theory" (or logic) of change and evolution, we are dealing with forms of reasoning which do not function by 'yes' and 'no', the principle of the excluded middle is clearly inoperative and in this thus totally different from formal logic. Ironically, Sanchez-Palencia although Spanish spent his entire academic career in the French CNRS in which Lupasco had started it.

3.6. Dialectical Logic

The paraconsistent logician Graham Priest [43] has pointed out that Hegel distinguished between dialectics and formal logic—which was for him the Aristotelian logic of his day. The law of non-contradiction holds in formal logic, but it is applicable without modification only in the limited domain of the static and changeless. In what is generally understood as a dialectical logic, which LIR superficially resembles, the law of non-contradiction fails. Subsequent developments of formal logic, starting with Frege and Russell, have forced Hegel's conception of contradiction to be rejected or interpreted non-literally. Neo-Hegelians have attempted to conserve *this* principle of contradiction by emphasizing the factor of time: A is not identical to A, because time has passed in which changes have occurred; contradictions take place one after the other, etc. Articles purporting to describe dialectical logics still appear. In one example, a relation is proposed with non-linear dynamics in which dialectical logic is enhanced by mathematical logic. Nevertheless, the question is not addressed, here and in Hegel, of what drives the change from thesis to antithesis to synthesis, that is, how any term cannot 'stand on its own' but 'goes over' into its opposite or contradiction. Russell demonstrated, before Lupasco, that Hegel's logic could be deconstructed because it still presupposed traditional Aristotelian logic, but not for this more important reason, namely, the absence of a grounding in physical reality.

Piaget, also, did not go beyond the standard Hegelian form of Marxist dialectical materialism. This correctly accords a central role to conflict and contradiction in the transformation of social realities. However as pointed out by Priest [44], and further discussed by Igamberdiev [10] and in this paper, Marxist dialectics fail to give an adequate account of the true contradictions involved in society: an inconsistent or paraconsistent logic is necessary for such an account, albeit in our view not sufficient. A logic of the LIR form seems required to characterize the emergence of new structures from real contradictions.

In the LIR view, cybernetic systems, natural or artificial, *are* dialectical, since each one involves an alteration, a perturbation by an antithetical contradictory process, followed by the return to the (state of) regulation required for the system to be "stable". In other words, a cybernetics alternately actualizes certain phenomena and potentializes the antagonistic, contradictory phenomena in consequence. It is an "oriented dialectical systematization of energetic events, inherent in the nature of energy" [45].

As will be discussed in more detail elsewhere, Logic in Reality differs from standard forms of thought in showing the importance of implication as a process in contrast to reliance on equations. The astrophysicist Stephen Hawking, for whom we have the greatest respect, once asked: "What if the universe and human beings were governed by the same equation?" We ask a somewhat different question: "What if the universe and human beings were governed by the same thing, but it is not an equation?" It is not an *inequality*, which to us a just an equation that does not admit it. It is something like an *inference*, but inferences are associated too strongly with language and its limitations. We have therefore pleaded in this paper for attention to be paid to the strength of *implication* in describing what governs the world. David Bohm [46] had a similar intuition in his proposal of an implicate as well as an explicate order in the universe.

3.7. Dialectics and Logic in Reality in Marx and Engels contra Hegel

With the basic concepts of Logic in Reality in hand, including its dialectical base, let us how it can be related to some dialectical aspects of the economic theories developed by Marx and Engels, as well as their roots in Hegel. The objectives of this paper do not include in-depth study of dialectics in economical and political theory and practice. In this Section, we simply look at some aspects of this field which can be usefully addressed by combining dialectics, as described in Section 2, with the perspective of Logic in Reality outlined here.

Engels saw dialectics as a fundamental method for gaining knowledge of nature, and as a thought process of involving a real opposition of ontological as well epistemological contraries, similar in appearance to that of Logic in Reality. Arthur [47] points to what we can see are critical weaknesses in Engels' dialectics. One was a tendency to rely on a compilation and classification of examples and another was the presentation of a triadic paradigm as the 'three laws' of dialectics', without a basis in science. To anticipate, we find exactly the same structural weaknesses in the epistemological triads of Charles Sanders Peirce that are considered the foundations of standard semiotics (next Section 4).

The point of departure for knowledge for Engels was the "qualitative aspect of things and phenomena and not their quantitative side". As Brenner has discussed elsewhere [1], for Engels "Dialectics is the science of universal interconnection, of which the main laws are the transformation of quality and quantity—mutual penetration of polar opposites and transformation into each other when carried to extremes—developments through contradiction or negation of negation—spiral form of development." Logic in Reality is a way of giving a physical picture of the transformation of the polar opposites operating in complex systems: it is never complete, and the extremes of 0 and 1 are never reached except in trivial cases. Contradiction or better countervalence inheres in physical processes, and negation of negation remains at the level of linguistic logic. The concept of spiral form of development is an absolutely essential one which Brenner described in detail as a consequence of Deacon's concepts, grounded in developmental biology, of "incomplete nature" [48] (cf. Section 5.1 below).

We look next at the discussion by Arthur [46] of the 'new dialectic'. For example, it is statements like those of Marx that correspond to our view of the ontological priority of real phenomena—the "logic of the body" rather than a "body for the logic", as echoed by the contemporary work of Lakoff [49]. 'Our' dialectics is systematic (with an internal dynamics) rather than historical (demonstrating temporal dynamics). However, it goes beyond the systematic dialectics of Marx in its necessary inclusion of insights from 20th Century physics. We can thus easily dismiss empty concepts in Hegel such the realm of logic being 'timeless'; only standard binary linguistic logics, including that of Hegel, are 'timeless'. Arthur further suggests that Engels applied what was basically a linear logic as he, as the majority of other thinkers, did not have room for the recursive aspects of real processes in the 'sinusoidal' movement from actuality to potentiality and vice versa. Marx's apodictic statements about the 'means of production employing workers' rather than the reverse need not be taken as dialectical contradictions in the LIR sense; the two opposites are in different linguistic domains, turning on two senses of 'employ'.

Logic in Reality thus provides a new, more physically acceptable view of neo-classical concepts in dialectics, for example that a logical progression is at the same time a retrogression. Instead of reading this as a succession of abstract categories, we consider it a description of the dialectical evolution of real processes, for which the simplest model would be graph with two mutually dependent time axes, and which, to quote Arthur, "has nothing in common with a vulgar evolutionism predicated on extrapolating an existent tendency", we add, unidirectional.

A feature of *Das Kapital* [50] is its exposition of the "reciprocal conditions inherent in a whole and not a quasi-historical (linear) development from primitive conditions to advanced ones". Logic in Reality establishes the applicable reciprocity in reality that gives this insight its scientific value. Where we disagree with the dialectical and logical formulations of Marx and Engels is in their essential use of standard categories as constitutive of a systematic dialectic and in their treatment of contradictories. Engels' famous 'three laws of dialectic', referred to earlier, such as the negation of the negation, are restatements of the laws of linguistic Aristotelian logic. The concept of the properties of capitalism as emergent from such laws is an epistemological one, and bourgeoisie and revolutionary proletariat are terms that lack sociological reality.

In LIR, these are not to be resolved in an eliminative sense, as it is from them that new concepts can emerge without requiring their total disappearance. However, Arthur suggests that one may "draw upon" Hegel's categories to show that anything and everything can become a bearer of value. Our position is no more or less than that anything and everything—that exists—is a bearer of value since it is a bearer of meaning. We discuss further aspects of meaning in Section 5.

It is however, exactly these 'classical' aspects of Hegel's logic that make it relevant to the ontological foundation of the capitalist system. As is becoming more and more obvious, the 'inhuman' aspects of the emergent contradictions in capitalism—the *absolute* dialectic of capital and its quasi-logical primitives (Arthur), are exactly where classical logic gives a reduced, binary picture of the world. It using the principles of LIR that we can accept, by giving a physical dialectical meaning to it, the statement that the separation of "quantity and quality from each other is not absolute".

We will not analyze in detail Hegel's views of Being and Nothing (to which we prefer the 'Nothingness' of Sartre). Heidegger's ontological *Dasein*, while not describing a sufficiently 'full' presence, approaches more closely the physics of LIR. Regarding Hegel's discussion of correlative pairs in thinking, the question cannot be of actualizing an 'inner unity' which does not exist.

Hegel's idealism and absolutism led him to make a further error in respect to necessity and contingency by leaving no place in his Essence-Logic for the latter, while ascribing 'Actuality' to the former [41]. "The sole aim of philosophical inquiry is to eliminate the contingent". What is false here, exactly as in the tychism of Peirce that sees chance as most fundamental, is the absolutism. It is the dialectic between chance and necessity, or determinism and indeterminism that is essential and it is Lupasco who deserves the credit for placing this dialectic on a basis in science.

There are serious limitations in Arthur's analysis of value and form, let alone dynamics and process, by his emphasis on linguistic concepts. In looking for the relation between form and content he mentions as an interesting case the "logical form of a proposition being independent of the variables in it. We speak here then of two sides indifferently united". We are far here indeed from the real world in which even propositions are not topic-neutral. We thus agree with Arthur that capital has a "certain conceptuality to it" in which a peculiar interpenetration of ideality and materiality is expressed as a contradiction between value and use value. That this leads to a "concrete mediatedness without ever reaching a final resolution" is far from concrete, however.

The Logic of the Concept defined by Hegel shows his struggle with the relations between individual and universal, continuity and identity. Marx used Hegel's terminology for his logic of capital: capital is a universal distinct from its moments (instantiations) while being at the same time continuous and identical with these moments. We will not attempt to discuss the historical validity of Marx's concept but only suggest that the dichotomies only make some sense when considered as partial and interactive in the LIR sense.

Marx's *critical* systematic dialectic is also closer to LIR than Hegel's *affirmative* systematic dialectic in that the contradictions inherent in capital are not eliminated or sublated but deepened or developed as it evolves. In the same Compendium, Carrera [51] cites Marx's aphorism that "philosophers (such as Hegel) have only interpreted the world in various ways; the point is to change it". Here Marx is being too classical: serious interpretations can also 'change the world', given a favorable terrain.

Both Hegel and Marx and their commentators suffered from reliance on epistemological readings of dialectics. Bellofiore [52] mentions the "inner connections of objects and concepts"; "capital as an invisible subject in a kind of perennial movement in a circle", moving[6] (?) from simple and abstract categories to more complex and concrete notions", with 'redefinition' of the categories." The problem in our terms is that both authors defined a category as a principle unifying different particulars, reconciling

---

[6]   The term 'moving' here raises many questions which it would be otiose to follow.

the universal and individuals by the principle of the complex unity of identity in difference. This fails (in LIR) by the impossibility of *complete* unity, when it incorporates the moment of difference. Stating that this demonstrates a *negation of the negation* has no meaning in an ontological sense. Although we thus have problems with the stages of Bellofiore's argument, we can agree with his conclusion, namely that we are dealing with two antagonistic ontologies, and a proper reading of *Capital* implies that Hegelian circular views about capitalism are false in practice.

In contrast to the heavily ideologically loaded and hence largely incorrect tenets of the economics of Hegel and Marx, the work of the Japanese economist Kôzô Uno [53] and his colleague and translator Thomas Sekine [54] illustrate the kinds of balanced, non-dogmatic positions that are possible in approach similar to ours. Uno argued that capitalism, throughout the 20th Century, must be studied at three distinct[7] levels of abstraction: (1) pure theory in the form of Hegelian dialectics; (2) concrete capitalist history, "which must be recounted with full empirical detail; and (3) at a mid-range level between the two, focused on capitalism's developmental stages". Obviously, it is this mid-range level that is most adequate for the understanding of complex real processes such as capitalism.

In his Appendix 2 to [54], Sekine characterizes the "true teaching of Uno's economics, which lies in his grasp of capitalism as something that exceeds the confines of formal (i.e., tautological logic)". Uno assigns proper ontological value to a capitalism whose nature cannot be concealed, and it thus can be seen to follow a logic that is independent of any ideology, Marxist or bourgeois supported by mathematical myths. A rational dialogue about the problems generated for both the society and the environment becomes possible.

Our approach remains beyond the criticism of dialectics because the dialectical concept in the form of a Lupascian synthesis avoids final truths that were claimed by Hegel, Marx and which, in their application to social reconstruction, resulted in totalitarian societies. In fact, Engels in his later years was ready to support the idea that social progress can be achieved by peaceful means, which include incremental legislative reforms in democratic societies as was further developed by his disciple Eduard Bernstein. The dialectical approach developed here is in concordance with the humanistic trend that can be dated back to a 1795 essay by Immanuel Kant "Perpetual Peace: A Philosophical Sketch". The main statement of Kant, which is truly dialectical, is that permanent peace is a goal which mankind can approach gradually in the course of its improvement. A conscious ethics commits us to act such that perpetual peace can be reached; this represents an example of the categorical imperative. With our new formulations of both dialectics and logic in hand, we now turn to their applications to aspects of meaning and its instantiation in the areas of semiotics, information and communication.

4. Semiotics and Semiosis: The Dialectics of Meaning

The views of dialectics and logic that we have developed can be used as tools to reconceptualize the areas of semiotics and semiosis. The key concept in semiotics has been the introduction of an ontological and epistemological role for the sign. Signs are elements or units of language that are assumed to convey meaning to living beings capable of receiving and interpreting them, that is, of evaluating their relevance to the on-going processes of survival and growth, physical and intellectual. Two names in the thought of the last 150 years are associated with semiotics as a discipline, Charles Sanders Peirce and Ferdinand de Saussure. The problem with these authors and their countless followers is that their theories are non- or anti-dialectic. Simply, contradiction is given a purely negative interpretation.

As soon as living beings appear in the universe, meaning appears as process aspects essential to their survival and reproduction. With the arrival of self-awareness of existence, say, at the level of human beings and cats, semiosis describes the active, ontological process of the *discovery* of meaning in

---

[7]   As usual in our critique, calling appealing to some abstract concept of distinctness adds nothing to the argument. It is neither necessary nor desirable to make absolute separations between dialectical theory and the necessarily historical developmental stages of a social phenomenon.

and as existence-as-process. It is creative, participatory and relational; semiosis carries an emotional stance or feeling toward existence and the self as part of that existence, which includes the capacity for its perception. It is a phenomenon within the scope of natural philosophy, with its own dynamic 'logic' of processes.

4.1. The Semiotics of Peirce

In contrast, the semiotics of Peirce is an epistemological doctrine describing the classification or categorization of the meanings or meaning-processes inherent in all existence. It is reductive and inert, since inserting the concept of sign or representation between a phenomenon and meaning adds no new meaning [55]. According to Peirce, semiotics is a logic and is logical, but only in the sense of standard, bi-valent or multi-valent linguistic logics. Peirce's famous 'three-tailed' graphs do nothing but provide a visual equivalent of bivalence.

For us, the discovery of new relations between categories is a semiotic process, but only because it takes place *outside* the epistemological categories. To say that it is 'Thirdness' is a tautology, and the entire structure of Peircean semiotics lies in the domain of non-natural philosophy.

Our concern with semiotics stems from the feeling not that it is wrong in some absolute sense, but that it addresses secondary aspects of the complex phenomenon of meaning and its exchange. The hermeneutic process adopted by Peirce illustrates this: he takes the extant domain and categorizes it without making an ontological commitment as to its basis, something which we consider an unavoidable necessity if the properties of existence are to be correctly represented.

In the realistic approach adopted in this paper, we suggest that the only possible hermeneutic process requires starting from a correct assignment of ontological priority. We have access to an historical emergence of knowledge, defined as a mental process for insuring the flourishing of living beings, as individuals and species. Since survival, if possible without pain, is the objective, knowledge of how to accomplish this is what is meaningful for the individual; it is constitutive of meaning.

Logic in Reality, which provides a rigorous framework for discussing presence and absence and their interactive relation, provides a non-linguistic definition of a sign as a perceived phenomenon in relation to meaning in the above sense. For a deer, the information received as the smell of a lion, or sight of its droppings, *are* the lion as a potentiality, in potential form. The deer, having interpreted these signs, undertakes actions to insure this potentiality does not become actualized nearby.

Peirce based his theory on a division of phenomena into categories of Firstness (possibility), Secondness (existence) and Thirdness (reality). The 'First' is a 'Sign' or 'Representamen' which is in a genuine triadic relation to a 'Second', called its 'Object' so as to be capable of determining a 'Third', its 'Interpretant' to assume the same triadic relation to its Object in which it stands itself to the same Object'. The term 'Sign' was used by Peirce to designate the irreducible relation between the three terms, irreducible in the sense that it is not decomposable into any simpler relation, such as some form of part-whole relation.

In our view, the Peircean categories invert the ontological priority in the universe and fail to add to knowledge or how to acquire it. The problem was succinctly addressed by Petre Petrov in his "Mixing signs and bones: John Deely's case for global semiosis", [56]. On p. 412 he writes: "The transformation of physical reality into objective sign reality is the same as the transition between Peirce's categories of Secondness (a binary relation of opposition, impact, cause and effect) and Thirdness (a triadic structure in which one item relates to another *for* yet another, the last one being the 'interpretant' of the dynamic between the first two).

The trick succeeds only because one has put the rabbit in the hat beforehand only to pull it out later. Secondness is already implicitly Thirdness, and the so-called physical relation is already implicitly objective" (italics ours).

When one tries to talk to Peirceans about possible overlaps between the categories when applied to real change, one is told "Oh yes, Peirce was aware of them and they exist". However, there seems to be no more grounding for any physical relationships in these than in the original categories.

Lupasco devoted major studies to application of his logic to living systems [45] as discussed elsewhere by Brenner. In current terms, his logic is consistent with a view that life emerges from the recursive processing possible in some still undefined proto-structures, and what can be called semiosis in higher animals from some further recursive processing of information within life's physico-chemical structures. Only physical processes are required for meaning; for us, they can be and **are** meaning. Signs are inert, *a posteriori* epistemological constructions without explanatory value. (In our approach, biological 'codes' are not signs. They are static scientific abstractions.) The concept of operators is useful here [57]; it is the chemical structure of nucleotides in the DNA that 'operates' the genetic process.) Let us now therefore leave the austere world of Peirce and his acolytes and discuss concepts of semiotics and semiosis from other logical and scientific standpoints

4.2. The Semiotics and Semiosis of Igamberdiev

From the earliest known examples of abstract thinking, humans faced the two types of paradoxes, one referred to as semantic and the other as kinematic. The first apparent formulation of the semantic liar (*pseudomenos*) paradox was that of Epimenides of Cnossos who was a semi-mythical thinker of the 7th or 6th century BC. Its later formulations belong to Eubulides (4th century BC), and in other cultural traditions to Bharthari (India, 5th century AD) and to Nasir al-Din al-Tusi (the Islamic tradition, Persia, 13th century), who identified the liar paradox as self-referential. This paradox substantiates the impossibility of non-contradictory application of binary logic to the process of signification (assigning of meaning).

The first known formulation of the kinematic type of paradox in the West was by Zeno of Elea, who was the disciple of Parmenides (5th century BC). They were formulated independently by the representatives of the School of Names (Míngjiā) in China (5th-3rd centuries BC) and by Nagarjuna in India (3rd century AD). The origin of kinematic paradoxes is in the impossibility of a non-contradictory formulation of an infinite space-time and to the incompleteness of any formal representation of it. In such representations, contradiction appears between the description of space consisting of an infinite number of points and the possibility of moving past them in finite intervals of time.

Over many centuries, the semantic and kinematic paradoxes were analyzed independently of one other. The main challenge was to try and unite them in a single dynamic process. This was, in particular, reflected in the designation of Cretans as liars in the Epimenides-Eubulides paradox in which the physical (kinematic) parameter (movement) was introduced. If we remain within the physical domain and observe a moving object (*e.g.*, Zeno's flying arrow) embedded in physical space, we can introduce a semantic constituent by assuming that the arrow, at a concrete finite moment of time, is present at a certain point of space, and at the same time is absent there, beyond the semantic field of the person who launches it (the bowman). The dynamic process thus includes the contradiction that is embedded into the moving system and can be fixed logically. Space-time is structured by physically defined parameters that appear to fulfil the function of avoiding the apparent inconsistency of its representation as infinitely divisible to allow movement in it. These parameters represent the fundamental constants of the physical world, and the physical complementarity which manifests itself, in particular, in the impossibility to define strictly the position and momentum of a particle simultaneously. This arises as the consequence of the process of representation into the represented (measured, observed), which continuously produces an infinite recursion. Quantum measurement appears as a semiotic process in which signification takes place continuously and in which the kinematic and semantic constituents are complementary and emerge in tandem from a still unknown *pleroma*.

The incompleteness theorems formulated by Kurt Gödel represent modern interpretations of the semantic paradox but at the same time they have a kinematic constituent which appears in the process of obtaining proofs. The claim that any sufficiently rich formal system is incomplete means that it contains statements which cannot be proved inside the system, but they can be enumerated (encoded) in the representation of the system. A system of axioms can never be based on itself since the statements about the system itself must be used in order to prove its consistency. The great invention of Gödel [58] in his first incompleteness theorem is that its proof is achieved via assigning statements about the whole system (corresponding to meta-mathematical statements) to the elements of the system. Signification is therefore attributed to the dynamic process of (epistemologically) developing the proof of incompleteness itself. Although this dynamic process is not analysed logically (or "dialectically"), it represents the necessary condition for obtaining the proof for the theorem. As shown by Igamberdiev earlier, *a* process of obtaining proof, which necessarily generates the encoding in the system, is continuously realized in living systems and their evolution [59]. However, the latter is not the same as proof of statements or their mathematical equivalents. At a higher level of reflexion, it is present as a dynamic process of evolution of social systems and civilizations [60]. A concept that attempts to explain biological encoding as a consequence of metabolic reflexive loops in the system, "closed to efficient causation" in the Aristotelian sense was developed by Robert Rosen [61], discussed further in Section 6. Its further development can be based on analysing the dynamics that results in generation and further complexification of these loops.

To understand this dynamic process, we need to analyze the course of obtaining the proof in Gödel's sense which will characterize the process itself and generate encoding in the system such that it becomes a semiotic entity (seme). During this process, growth of complexity takes place and the system acquires *information* about itself and about the external (to it) reality. In other words, Gödel contributed much more for understanding semiosis than Peirce or Saussure because he explained the necessity of encoding as a required step in the signification process. This process represents a dialectical discourse of the system with other organized systems embedded in the systems reality external to them... The complementarity between the formal Gödel system and Gödel numbering is based on the separation between the system itself and its embedding that represents its encoding. It follows or tracks the growth of complexity via the dynamics that is inherent to Gödel's self-reflexive loop and includes a certain freedom as the encoding (Gödel's enumeration) can be reached by many alternative, i.e., complementary, ways. Thus the growth of complexity is based on the contradiction between a system and its representation (including its internal "encoded" representation), generating new levels of "gradation" of organizational structure, and includes the relational abolition (negation) of previous steps. Some arrangements that were actualized in previous steps emerge later but at a higher level of complexity, as suggested by the Axiom of Emergence in LIR. These moments of the dialectical discourse loosely correspond to "three laws of dialectics" formulated by Engels. This dynamic process was defined by Heraclitus as the self-growing Logos (*logos heauton auxon*).

When Kurt Gödel enumerated metamathematical statements about the system by encoding them within the system, he incorporated at the same time the paradox in the nature of the dynamical system. The next task is to describe in logical terms the procedure of the proof of incompleteness itself. This introduces another paradox: to describe the origin of logic by logical means and the origin of computation by computable means. This means that we enter into a discourse between our own logic and the internal logic of the system with which we interact and which we aim to describe. The nature of this paradox has not been fully analyzed in human thought, except in Plato's dialogue "Parmenides" and perhaps a few other philosophical works. It is very easy to slip from the dialectical discourse to the concept of final truth. This was not avoided by Plato himself (finally in his "Laws") and by Hegel and Marx who abandoned their own well-developed discourse for such a "final synthesis". This synthesis, in fact, rejects the very nature of dialectics and results in total failure for the conceptual field and in social practice.

For resolving the paradox of describing the origin of logic by logical means, it is important to consider Friedrich Wilhelm Joseph Schelling as an essential figure in natural philosophy. According to Schelling's views, since the coming into being is not a finished entity but yet becoming and always contingent, existence and movement cannot be a logical category [62]. For the purpose of describing the process of actualization, we need an instrument that can in some way overcome the frames of purely logical description of the process of coming into being, and dialectics represents a substitution for descriptions using classical logic. Arran Gare [63], analysing Schelling's views, comes to the conclusion that Schelling developed a theory of emergence and a new concept of life relevant to current theoretical biology. This theory is grasped in Logic in Reality in the Principle of Dynamic Opposition (PDO) operative at the most fundamental physical level as well as all higher levels of reality. The PDO and contradiction or counteraction, and a robust notion of potentiality, is required for understanding the relation between substances, events and processes. (We note in this connection the debate in Buddhist logic on whether dynamic (al) opposition was or was not real ([38], pp. 404–406), involving, like LIR, an included middle.

The concept of emergence introduced by Schelling was further developed by Alfred North Whitehead [64]. Following Whitehead, it is only events that are the actual entities of the physical world which need to exist in order to have a physics. This leads to a new idea of physical substances, that of a distribution of potentialities, "powers" or propensities, that are "reasonable consequences of a theory of processes" as Thompson [65] suggested. He talks of eternal objects with only a 'pure potential' for ingression into reality in which potentiality is realized (actualized). However what Whitehead designates as Categories of Explanation include highly relevant statements of the ontological foundations of Logic in Reality as a logic of process [64]. Thus

(i)   (That) the actual world is a process and (that) the process is the becoming of actual entities.

(ii) (That) *how* an actual entity becomes constitutes *what* that actual entity *is*. . . . Its being is constituted by its becoming. This is the principle of process.

We do not hesitate to characterize this aspect of Whitehead's philosophy as natural: "For you cannot abstract the universe from any entity, actual or non-actual, so as to consider that entity in complete isolation." Whitehead contrasts the philosophy of organism, basically what we discuss in this paper, with that of propositions as abstractions. It is of interest is that Whitehead has no difficulty in integrating being and becoming without conflating them. Whitehead integrates both physical and philosophical senses of 'potential difference'. The further differentiation of processes by Whitehead into macroscopic and microscopic, efficient and teleological, complete and incomplete and their comparison with the principles of Lupasco will be discussed elsewhere.

Another important philosophical concept of organizational evolution of matter is tektology of Alexander Bogdanov [66]. It was analyzed recently by Arran Gare [67]. In this concept Bogdanov put process and organization in the center of perpetually evolving reality seen as an organizational process of the Universe. Science itself, according to Bogdanov, is comprehensible as a development within and of nature. Tektology claims that processes are the primary reality rather than things or substances and their attributes, which assumes that a non-formal language is more appropriate for characterising the basic characteristics of nature, and mathematics should be seen as having a derivative status [68].

The aim of modern synthesis in science is to explain the phenomenon of complexification which lies at the foundation of evolution of the Universe [69,70], in evolution and morphogenesis of living systems [71–73] and in social progress [74,75]. For this purpose, the dialectical ideas of Plato and Aristotle revived in modern times by Schelling, Whitehead, Bogdanov and Lupasco represent an essential alternative to the positivist interpretation of reality that dominated science for many years.

4.3. Epistemic Perspectives and Logic in Reality

The perspicacious reader will have noticed that, although we claimed at the outset to focus on the ontological entities of natural philosophy, we have introduced *three* concepts in the discussion that are primarily epistemological: (1) the *clinamen* in Section 2; (2) the ontolon above, which retains substantial epistemic character in its relation to the epistemon, and (3) essential properties of quantum entities that are outside standard quantum physics as generally understood.

We justify the inclusion of these perspectives, as we do that of Peirce, for their hermeneutic value. They can provide additional insight once it is established and clearly understood that there is no direct physical evidence for the existence of the respective entities *qua* entity. In fact, we have returned in restated terms to the dialectics of appearance and reality. The novel feature that we are able to introduce into this dialectics is the availability, offered by LIR, of a dynamic relation between the two—no appearance that is not also partly reality and vice versa, sometimes changing in proportion with time. Meaning inheres, implicitly and explicitly, throughout the domain.

Let us now examine these considerations in the newly available language of information science and philosophy.

5. Information and Communication: The Dialectics of Meaning (2)

In Section 4, we saw an example of a field of which both formal and non-formal descriptions are possible. Formal approaches are not necessarily incorrect, but they are when they equate to the use of bivalent or multivalent linguistic logics and standard category theory which in our view have limited explanatory value. Non-formal or partly formal approaches have been much less developed, but they offer interpretations of phenomena in terms of dialectics and a non-standard logic of processes, Logic in Reality (LIR), as outlined in Sections 2 and 3.

The two lines of thought are present throughout the human and social sciences and philosophy. Again, the second group has received less attention as being allegedly unscientific and non-rigorous, as most of the time it is. In this Section 5 and the next Section 6, we will address the fields of information and communication and *their* relation to meaning. We will provide a summary of the standard well-known interpretations which we will describe, for convenience, as non-dialectical (ND) and then of ours based on the parallel use of dialectics and LIR.

Also in previous Sections, we have referred to both meaning and information separately, pointing out some of the difficulties with standard conceptions of meaning in relation to semiotics or semiosis. In this Section we study what can be gained in the understanding of both concepts by looking more closely at the dialectics and logic of the relation between them. Even placing them in conjunction has significant consequences—in what sense are meaning and information the same or different?

Many authors have noted the complexity of information and the difficulty of giving a 'single, clear' definition of it. Attempts to do so are typical of standard substance ontologies, where hard definitions are automatically given preference. The failure of such attempts suggests that major categorial errors are being made. We therefore make the following lapidary statement which we will try to justify in what follows:

Meaningful information is reality in potential form.

It is derived from the Lupasco/LIR conception of consciousness outlined in [76] which basically looks at the real dialectical interactions between internal and external, better internalizing and externalizing processes as they move between potentiality and actuality.

In order to understand how our approach to logic and dialectics adds value to the discussion of meaning, information and communication and their interrelation, we need to summarize briefly in this Section the major developments in these fields made in the 20th Century and the last twenty years, both in the West and in China. The concomitant formulation of theories of information in which it was related to the clearly developing information society resulted in a new rationale and methodology for information studies. From the point of view of this paper, the overall movement was from non-dialectical mathematical theories to dynamic ones, including critical aspects from biological science is to be welcomed. Yet each of the kinds of theories outlined in the Sections 5.1–5.4 suffers in some way from the absence of explicit recognition of the dialectical/logical principles underlying the operation of 'Information in Reality' [30].

5.1. Communication Theory—Information Science—Meaningful Information

The concept of information as an 'immaterial' substance, involving meaning in some way, has been available since antiquity (cf. Capurro [77]). It is fascinating to note that the nature of information, often viewed today as 'just' a philosophical question, in fact requires the best available concepts of physical science of the origin and fundamental structure of the universe to be correctly addressed. The modern origin of a theory of information is often ascribed to the work of Shannon on theories of communication, in which information is related to the removal of redundancy in a formal manner. This approach was specifically intended *not* to address or define meaning, but it nevertheless gave rise to many other formal theories of which the most modern and comprehensive is that of the mathematician Mark Burgin [78].

In the 1950's however, the discussion of information was almost fatally polarized by the statement of other mathematicians, Norbert Wiener and his follower John Wheeler, to the effect that information was not matter or energy. This statement gives the sense that there is something truly 'different' about information, but scarcely gives an indication of what it might be. Such theories are not wrong but they are incomplete in that the scientific and logical origin of the dialectical, that is, in our terms real physical interactions described is not specified. Faced with the multitude of theories of information of all kinds, Brenner gave his 'personal synthesis' [79] of a several competing theories, from which a few key benchmarks for philosophy can nonetheless be taken as discussed further in this Section.

Given their origin in communication science and its significant dependence on technology, it is not surprising that theories of information emphasize formal concepts and standard logics. From the human, *natural* philosophical standpoint—we do not separate these—communication as the transfer of information is a human activity, and its nature and changes must follow the rules and logic of the latter.

5.2. Geometry/Position or Energy/Force

In seeking the grounding for philosophy and knowledge, one is confronted by at least two age-old, unresolved arguments. One is whether space, defined by position and geometry or energy, defined by movement and change is more fundamental—has ontological priority—in the universe. Another is whether matter is somehow constituted by our consciousness of it, the doctrine of anti-realism, or consciousness evolved from or together with matter. Those who prefer the first elements of these two pairs, including the major figures of Norbert Wiener and John Wheeler, also favor the position that information is *sui generis*, neither matter nor energy, present in some way in the form of digital bits, the so-called It (things, processes) from Bit position.

The proposal of Logic in Reality outlined above and in [80] is that what emerges from the still unknown ground of the universe, position and energy, or statics and dynamics, can be rigorously described in dialectical terms as two different, opposing but non-separable aspects of that ground, a position similar to that taken by Diaz Nafria and Zimmermann [81]. Digital bits, now part of knowledge are an emergent epistemological phenomenon—the 'Bit-from-It' position, and the overall situation can be described by an 'onto-epistemology', whose components are 'ontolons' and 'epistemons'.

If as argued energy is fundamental, so in our view are its properties, in any real system, of actuality in a non-separable relation to potentiality. Change is a consequence of our existence within an overall energy gradient moving locally from higher to lower grade forms, finally to heat (low energy photons). The movement of elements itself is sinusoidal, from predominantly actual to predominantly potential and the reverse. No element goes to the absolute, idealized limit of 100% one or the other, except in trivial cases in which there is no interaction between the initial and final states, as in snooker balls falling into pockets. The hypothetical pre-thermodynamic universe is supposed to be something like a quantum vacuum, with elements moving into and out of existence, without reference to time. As attempts to show a direct coupling of such a world with ours, we feel that they are not required for our subsequent analysis. The desire to demonstrate such coupling must be placed in the same domain as that of religious beliefs which by definition cannot be the subjects of scientific proof, except of their *own* existence.

As proposed by Lupasco and discussed by Brenner in [1], whether a dialectics of energy and position, or of consciousness and non-consciousness or of a polarized position of absolute opposites is preferred is a question of individual psychology. Lupasco saw individual human beings as instantiating change and hence preferring the stability conferred by belief in identities, stasis, rather than the dynamics of diversities. This dialectics, the dualistic oppositional properties of energy is reflected in the properties of information. Information and energy are both the same *and* different, as the focus of the mind moves from one aspect to the other.

The major difference is, of course, that information is alleged to be and convey meaning, and the emergence of meaning from a meaningless substrate has been impossible to explain. In our view, the question of the emergence of meaning is equivalent to that of consciousness as a higher order of mental processing by animals capable of that processing, related to their survival. We admit the tautology here, but claim that the fact that no explanation of the emergence of meaning exists does not mean that the ascription of dialectical properties to information is invalidated.

5.3. Why Information is Enough

The cybersemiotic approach of Søren Brier is outlined in his major book [82], *Cybersemiotics*. The title reflects is program of defining a new semiotics that in his view emerges from the computational processing of knowledge. The subtitle is *Why Information is not Enough* and refers to the limitations of standard logical and computational theories of information. Brier refers here to the lack of an adequate rigorous framework for philosophy, now including phenomenology, in an adequate relation to the natural sciences.

As Brenner has noted elsewhere [1,30,55], Brier among other semioticians have based their underlying world view on proposals of Charles Sanders Peirce for logic and semiotics, in fact of a theory that implies their equivalence. Peirce made a classification of phenomena into those instantiating categorial properties of Firstness, Secondness and Thirdness, roughly related to the degree of interaction present. The problem for us is that the properties selected have, by Peirce's own admission, no grounding in any physical reality. They thus rejoin the body of knowledge referred to in Section 5.2 as anti-realistic. The semiotics of Peirce, and accordingly of Brier, is a description of the world, including living systems, in terms of 'signs' to which human beings can give an interpretation = meaning. Biosemiotics has become a highly active field of inquiry designed to explicate the role of complex signs in knowledge.

In our view, the approach of Brier and other semioticians displaces the emphasis from the ontology of the world to its epistemology without giving an adequate reason or substitute for it. Peirce did not claim that his system was based on any ontological commitment as to the ground of his signs, but that did not prevent him or his followers from assigning major hermeneutic value to them. We will return elsewhere in this paper to a discussion of phenomenology as such. We simply restate here the advantages of the LIR physical view of the energetic aspects of the origin and propagation of information, without reference to signs. Information inheres in both the physical and epistemological evolution of real processes and hence serves as a concept unifying the physics, biology and neuroscience of mind.

The somewhat apodictic statement by Lupasco cited above that physical processes, and hence their informational content *are* meaning is supported by the Kaufmann-Logan notion of biotic or instructional information [83]. It is also dualistic approach in which intrinsically meaningless Shannon information, the lowest level that characterizes a physical system without the self-reference present in living systems, is contrasted with biological information which always entails meaning. It emerges from the material-energetic structure of the latter in a process of which the 'structure' is a dynamic part. As originally emphasized by Lupasco [84] "everything is structure" or better a *structuring* (*structuration*), and Logic in Reality describes the evolution of these structurings.

5.4. Information in Presence-Absence Dualism

The basic philosophical position of Logic in Reality requires a dynamic interaction between opposites at and in respect to all levels of metaphysics. We have seen above the dialectics and logic of the relation between the degree of actuality and potentiality of opposing process elements, as well as the new entities which can emerge from them (included middle or third term states—'T-states'). A key pair of metaphysical opposites to which we ascribe a physical dialectical meaning are presence and absence. We have noted that LIR ascribes value, in ways that are more explicitly scientific as well as logical, to concepts generally considered negative—contradiction, inconsistency and vagueness. This approach is echoed by the work of the biologist Terrence Deacon of a metaphysics of incompleteness [85]. Deacon defined what is missing from theories of information in a paper with that title [86] and that this absence is an essential part of its content. For us, also, information refers to something that is not totally present now, or is not present yet.

We will not repeat here in detail Deacon's key concept that information is a relational property of systems that emerges from constraints of signal probability, discussed by Shannon, signal generation, by Boltzmann and those required for an apparently teleological dynamics, essentially those of Darwin in regard to evolution. We refer the reader to Deacon's major work, *Incomplete Nature* [48]. The three properties reflect three levels of entropy reduction that is an informational 'architecture' of recursivity. There is a rough parallel here to the notions of syntax, semantics and pragmatics, here, the pragmatics of Darwinian survival. The hierarchy of levels is that of data, content and significance, but the principles of LIR permit an interaction between them rather than the absence of interaction as between quantum and non-quantum levels.

At the cognitive level, according to Lupasco Principle of Dynamic Opposition, presence and absence correspond to consciousness and unconsciousness and actuality and potentiality in a non-intuitive contradictorial manner. Forgetting as also an active process, and in the complex dynamics of mind, a 'fact' or concept that was present and then relegated to the unconscious is an actuality with a potential for being recalled into consciousness.

5.5. Information and the Laws of Thermodynamics

In a general way, the discussion of complex real processes still suffers from the lack of an appropriate language which takes into account both intrinsic and relational properties of a system at the same time. Deacon defines constraints as relational properties, but LIR amplifies this by the rules for their evolution of the potential as well as actual aspects. Ulanowicz [87] has made an extension of the Deacon approach by connecting concepts of entropy to the Third as well as the Second Law of Thermodynamics, in order to define entropy in a relation to a degree of system constraint (actuality) and its conjugate state of residual freedom interpretation of the Third Law, deliver meaning. In a picture of entropy and information whose terms are always relative implies that like those of quantum physics, they do not commute [88].

In several papers, Igamberdiev further develops the idea that the Third Law of Thermodynamics is more important for understanding life than the Second Law which is considered the basis of Prigogine's dissipative structures. The Third Law establishes the reference state with the lowest entropy in relation to which the order (described as information) can be referred [89]. This state, according to the Third Law, is achieved at the temperature of absolute zero. However, living systems operate at temperatures near 300°K, in fact far from this reference state. It has been suggested that they maintain a long-lived cold decoherence-free internal state (called the internal quantum state), within macromolecular structures which is achieved by applying error-correction commands to the internal state and by screening it from thermal fluctuations [90,91]. Iosif Rapoport was the first who suggested to describe the stability of genetic structures by introduction of special thermodynamic principles explaining the maintenance of low entropy in living systems [92]. Mental processes could be associated with such long-lived internal states maintained within the nervous system [93]. This could be sufficient for Lupascian T-states, and a corresponding non-Boolean logic, as a possible physical precondition.

In our picture, the difference between these domains and those in which classical (Boolean) logic and the law of the excluded third operate is determined by the degree of interactivity and sinusoidal movement between them (primarily actual to primarily potential and vice versa, alternately and reciprocally), involving communication between actual and potential states on a macroscopic level. Stated in this way, the Principle of Dynamic Opposition (cf. Section 3) goes beyond thermodynamic principles. It requires implications in addition to mathematics for description of its operation in complex processes, given their non-Boolean evolution. The operation of the PDO thus functionally replaces the notion of a 'temperature' in metabolic cycles. According to Nicolescu [94], the Principle of Dynamic Opposition is not and should not be described as a thermodynamic principle. The thermodynamics of energy underlies all phenomena in opposition, but it does not characterize all of them. Ontolons, as described above in our picture, are constituted by predominantly actual states, but these are not independent of potential states. Together they lead to the emergence of new entities (T-states). Further complexification of ontolons at the interpersonal level generates perpetually evolving socially organized structures [75,95]. In the next Sections 5.6 and 5.7 we will summarize the approach to the science and philosophy of information coming from the philosophical side, which we immediately wish to characterize as part of natural philosophy.

5.6. Wu Kun and the Metaphilosophy of Information

The 35+ years of pioneering work of the Chinese information scientist and philosopher Wu Kun has inspired the following discussion. Wu's innovation in philosophy is that information is not immediately given as matter or energy, even though always requiring them, implies an emerging crisis in philosophy. Information is required as an additional philosophical category, and Wu redefined the philosophy of information as a *metaphilosophy* [96]. The implications of this change are only beginning to be realized.

Through papers in English by Wu and jointly by Wu and Brenner, including some in this journal, many of Wu's conceptions are now broadly available. A key aspect of this approach is that it *defines* a *stance*, the Informational Stance. Wu and Brenner consider that the "opposites" in information are not captured by the classical concept of a classical, static "unity of opposites", but by the dialectical interaction of opposites, classified by Wu based on his general philosophy of natural ontological levels that captures the essence of the properties of information. The resulting doctrine of objective information, subjective information and human information in society constitutes Wu's information theory and establishes it as a unified philosophical foundation for information science.

The Informational Stance, is a philosophical position and attitude that is most appropriate for, and above all not separated nor isolated from, the emerging science and philosophy of information itself. The Informational Stance is a more 'active' formulation of Wu's concept of Informational Thinking, which offers an alternative to standard 'Systems Thinking' in which most standard views of logic are retained. The Informational Stance is an attitude that requires attention to the informational aspects of complex processes as a methodological necessity, starting from the level of an existence theory for information and a methodology for its investigation. Especially, the Informational Stance supports and generalizes the recent work of leaders in the area of information ethics, including Floridi, Capurro and Wu himself, grounding the attribution of ethical value to all existence in informational terms. Wu's Philosophy of Information combined with LIR yields a *philosophical* structure of information that is compatible with its dynamic physical and logical structure and has no obvious direct precursor, either in or outside of the field of information.

5.7. Meaning and the Convergence of the Science and Philosophy of Information

Wu's definition of the role of information in philosophy is the critical *first step* in the characterization of the complex phenomenon of information and information processes. Further, Wu's classification provides a basis for an understanding of a key current development, the convergence of Information Science and the Philosophy of Information as the precursor of en emergent Unified Science of Information [97]. This convergence is obviously not intended to imply an 'end' to philosophy or its conflation with science. Philosophy will continue to explore issues that arise, in particular, in relation to language and knowledge in their aspects as unique cognitive products of the human condition, with a substantial abstract content. But the question of the relation of that condition to the rest of the world logically requires retaining the scientific properties of that world to insure the validity of the comparison.

There is thus a set of new and unique relationships that are developing between the classical disciplines of science and philosophy as a consequence of new understandings of the science and philosophy of information. The overall movement is that of a philosophization of science and a scientification of philosophy leading to their convergence. However, in this paper and elsewhere, we use the term Unified Science of Information. This is not strictly accurate, as our convergent theory includes the Philosophy of Information as a proper part, without conflation. Wu and Brenner therefore propose, despite its awkwardness, the term Unified Science-Philosophy of Information (USPI) as the best possible description of the field of endeavor. We thus believe we are witnessing the emergence of a new system of science, a metascience in a complex, dynamic reciprocity with philosophy that amounts to a paradigmatic revolution in thought.

In a paper in *Information* [98], Wu describes the current situation as follows. We cite this passage *in extenso* as it is a textbook example of the kind of new paradigm we referred to at the objective of this paper. "As a result of establishing the fundamental role of information in the existential domain, the Philosophy of Information provides a kind of dual-existential and dual-evolutionary theory of matter and information which describes information as a general phenomenon existing in everything in the cosmos. This leads to the acknowledgement of the dual dimension of matter and information in all forms of research. Because the lack of an informational dimension in traditional philosophy and science, it is necessary to transform them completely to take into account the new scientific paradigm provided by the current Science and Philosophy of Information. By means of that transformation, all scientific and philosophical domains become involved in an integrating, developing trend of paradigm transformation, which Wu and Brenner has called the "informational scientification of science" [99].

The current interaction and convergence of the Science and Philosophy of Information represents a fundamental and basic path for the development of scientific and philosophical knowledge. This "philosophization of science" and "scientification of philosophy", anticipated in the progression from ancient philosophy to modern science and philosophy, now represents a completely new way of thinking that is that distinct from that in the contemporary Western philosophy of consciousness. It resists an absolute separation between science and philosophy and establishes interactive, mutually defining feedback loops between science and philosophy which emphasizes their interrelation.

Given the properties of information outlined above, we claim that the relationship between its science and its philosophy is one of non-separability, leading to the convergence described. We assume that the Philosophy of Science is not identical to the Science of Philosophy, which remains to be defined. However, if as Wu and we suggest science and philosophy are converging, under the impact of the science and philosophy of information, the two cannot be considered as totally independent, separated or separable. The emergence of a revolution in philosophy, as suggested in [100] implies a revolution in the philosophical perception of science. In this paper, we have defined the parameters of and paradigm applicable of the Revolution in Philosophy and its implications for the Philosophy of Science. We see the two as a pair of doctrines in opposition in the sense of Lupasco. One may talk about a "New Kind of Philosophy" by analogy with the "New Kind of Science" proposed by Wolfram [100].

5.8. Logic in Reality, Meaning and Information

We have shown above that meaning and information are not identical but also not separable at the cognitive level which is the one of interest here. They follow the logic of real processes that is the thread running throughout this entire current paper. It is thus an integral part of Philosophy in Reality and of our proposed synthesis with the Philosophy of Information of Wu Kun. In the changes in stance or perspective, dialectical movement **is** both epistemological and ontological. Knowing not is not totally separate from the Known, what we know. It is true that the "map is not the not territory", but stated in this simplistic way, the relation restored or recovered by LIR is obscured.

The hierarchy for existence and essence established by Heidegger and by Capurro [101] and its formalism (*_is_*, *_as_*) can be placed into correspondence with a logical-physical theory of real processes, that of Lupasco and Brenner. The two 'languages' are linked by a concept of a dynamic relation between potentiality and actuality in both the physical and philosophical senses. The first consequence of this approach for semiotics is that one can clearly differentiate between the description of language as 1)) becoming, being-in-the-process of becoming meaning, involving real potentialities and actualities and their mutual conversion into one another and 2) as being - a static, binary phenomenon, in which philosophical potentiality inheres. McLuhan's apodictic statement that "The medium is the message" is only meaningful if the '_is_' is at the same time an '_as_', whose ontological sense is that of real processes of generating, sending, receiving and understanding messages as in the Angeletics of Capurro (see next Section). The current usage of media as a singular noun is incorrect from the standpoint of Latin grammar in referring to several kinds of medium—press, TV, cinema, etc. Oddly, it is more correct if 'media' is taken to refer to the complex dynamic properties of the messaging process, "the media is the message". The process and the process_as_meaning then have the proper ontological relation and value. A good example of our current approach is in its reading of the 'liar paradox', which was referred to and analyzed in Section 4.2. In the usual linguistic perspective applied, it is an epistemological dead end, oscillating between ideal limits of one or the other of two idealized alternatives. In the original perspective of one of us (AUI), the system subdivides into levels, with two actors. In the epistemological mode, Epimenides and the Cretans are separated by non-existence: Epimenides is a signifier, and the Cretans are the signified. The standard logical contradiction, which appears when we realize this one-level formalization of this system, is to be expected in this mode.

From our newer ontological perspective, the actors are the same but the relation is logical and contradictorial in the sense of Logic in Reality. The following consequences can be deduced from this consideration. Epimenides as an individual can be regarded as an element of the set which signifies this set, having *discovered* it in the sense of Capurro. The set (the society of Cretans) acquires its own dynamics having acquired the property of being 'liars' in conflict (antagonism) with, but not totally separate from, Epimenides. Different possibilities for the dynamic behavior of a system arise from this. What is important to note is that the 'dynamic behavior' with which we are concerned is that of real persons, individuals and groups, and their relations, dependent on their deep psychology.

On this basis we go to the next level of complexity which is that of meaning and information in human communication, outlined in Section 5.9 below. At this level, we will pay close attention to the concept, also due to Capurro, of "angeletics" or messaging theory as an extension of information theory as such.

5.9. Progress in Communication Theory

We have indicated in Section 5 some aspects of recent rapid developments in the science, logic and philosophy of information in relation to meaning. In particular, we have mentioned the dialectical convergence of science and philosophy under the influence of the philosophy of information. The convergence is no more functional than in the field of communication. We have also mentioned briefly their origin in the last Century in the science and technology of communication and their enormous increase with the advent of computer technology. Progress in communications is a part of the current digital ontological paradigm identified by Capurro in numerous publications in which, also, attention is called to their social consequences [102].

From our standpoint, progress in communication is by no means univocal. In the case of natural languages, for example, the basis for human communication, their evolution in the direction of increased efficiency and reduced redundancy is clear. The latter is off-set in many instances by context, as in Mandarin Chinese, where a single tone-symbol may have a dozen meanings, but this may be an exception. That such developments are accompanied by an impoverishment in the quality and depth of language and communication is also clear. Routine errors in English are made and become codified in defining media, e.g., CNN. The loss of redundancy in individual languages is paralleled by the disappearance of entire 'native' language groups. The destruction of the languages of indigenous populations employed was part of the political strategy of the Catholic Church in Mesoamerica, the conquest of the North American continent by Europeans and is part of that strategy in Asia and Africa today. The society is impoverished by such losses as much as it is by the loss of animal species, such as the moa[8], literally eaten to death in Australia in the 19th Century.

Under these circumstances one can only welcome movements in the direction of the saving of surviving languages and the recomplexification of our own. Such a process cannot be formal and artificial, but we propose what is essentially a third case of recovery, by the reintroduction of references to classical portions of our cultural heritage, not classicism for its own sake. We are thus proposing a rehumanization and de-digitalization of communication theory which is not intended to eliminate digital perspectives but to provide some balance to them. Capurro correctly points to the prevalence of a 'Digital Ontology' today [101], but this does not mean that a non-digital ontology is not active, admittedly to a lesser degree.

5.10. Messaging Theory (Angeletics)

Following McLuhan, further emphasis has been placed on communication as the exchanges of messages—'messaging' involving the triple of message, its sender and its receiver. Capurro coined the term 'angeletics' to replace that of information for the description of the exchanges of messages between human beings, going back to the ancient Greek word for messenger *angellos*. The choice of the term Angeletics for the study of messages and messaging is to signal the use of a philosophical framework, closest to that of Heidegger, which is ultimately based on Being of which the irreducible uniqueness of the individual human consciousness is a part. A major discussion by Capurro and others of the concept of Angeletics is given in [103].

The word message is derived from the Latin *mittere*, to send. For us, this describes a 'packaged' portion of a complex process of information in movement that can be also designated as an ontolon. This is consistent with the concept of Holgate [104] that science should continue "the epistemological revolution of our time" and extend relational, that is, ontological concepts to every field of science and the mind (relativity, pluralism, polarities, information exchange, etc.).

Communication involving the sending and reception of a message is a creative act, an energetic process that involves the exchange of energy, overcoming resistance to doing nothing, or not sending any message. Messaging theory or Angeletics must reflect the creative, value-laden characteristics of communication, their reference to the physical survival of the receiver, or more indirectly to his/her mental and spiritual well-being. It is easy to show a central role in philosophy of messaging in a historical perspective and more recently in the ontology of Heidegger. It is this central functional role of philosophy in messaging theory that calls for definition of a new specific field for the clarification of residual problems at the interface of the domains of messages, communication and information, in which the notion of Being plays a central role.

Capurro stated in [101] that the ontic difference between a sender a messenger and a receiver as separate entities presupposes the original unity of Being as sender, the world as message and humans as messengers. We can distinguish this original unity analytically, but we must be aware that any ontic separation (at whatever level of reality and concerning whatever kinds of beings) presupposes our being-in-the-world (to use Heidegger's formula). At the risk of overdoing the form but to make the point, we may use the expression Being in Reality in place of this formula. We propose a synthetic *non*-separation of subject and object and state grounded on the original relationship between man (and humans) and world. In an original angeletic perception, whatever we perceive AS being this or that and on whatever kind of relation (causal or not, etc.) is achieved on the basis of our being originally open to the message of the world that we process, at a higher level, as Being. One of the

---

[8]   "Can't get 'em, they've et' 'em,, they're gone and there ain't no moa." Australian popular song.

problems posed by discussion of Being in non-transcendental terms is whether one can talk also about it in terms of proper parts, as one can for ontic and epistemic phenomena, using the terms of ontolon and epistemon respectively. This problem is discussed in the next Section 6.

6. Forms of Reality and Existence

The dialectical-logical approach that we have adopted in this paper offers support for a new view of the operation of thought and other real processes at a further complex level. We do not consider this another level of abstraction, in the sense of Floridi [105] dealing with objects 'abstracted' from reality, but in the sense of sets of dynamic moving elements, parts or forms of the processes involved, potentia and actual. It is useful for this purpose to maintain the two domains of ontology and epistemology, where for the latter the concept of 'epistemons' has been proposed by James Barham [106,107]. The 'pieces' of the former will be called 'ontolons'. These two sets of entities could be supplemented by a third referring to the domain of standard philosophy, that of being[9], but this issue is outside the scope of this paper.

The relation of entities to disciplines is thus as follows:

1. Becoming/Change   Ontology   Ontolons
2. Knowing   Epistemology   Epistemons

No absolute separation should be taken to exist between the domains. They interact dialectically and logically in our LIR.

The terms of ontolons and epistemons are neologisms, but their derivation from Greek for 'being' and 'knowing' anchors them in a long tradition. Capurro [108] has made the somewhat easier connection of epistemon to 'what can be known', and has related it to Husserl's 'noesis'. This for us implies a process or stream of 'data', better, segments of processes. Ontolon is more complex, as it refers to both the immanent and transcendent aspects of 'being', the former in the English verbal sense. Our logical approach, however, suggests that these aspects are not and do not need to be totally disjunct. This is true in particular in relation to beings who, as we do, have the capacity of self-reference and an 'understanding' of our position, as in Heidegger. In Lupascian terms, we *are* both immanent and *transcending*.

Capurro states that, for Heidegger, ontology and phenomenology are not two distinct philosophical disciplines. These terms characterize philosophy itself with regard to its object and its way of treating that object. Philosophy is universal phenomenological ontology, and takes its departure from the hermeneutic of *Dasein*, which, as an entity—'analytic' or what we call proper part of existence—establishes a guide-line for all philosophical inquiry "at the point where it *arises* and to which it *returns.*" LIR provides the framework for the parallel interactive interpretation of ontology and phenomenology.

In contemporary philosophy, phenomenology thus occupies a strange position: on the one hand it seems to place major ontological value on appearance while at the same time denying access to it by science. Phenomenology could thus have the not very felicitous designation of anti-scientific realism.

6.1. Epistemons and Ontolons

Igamberdiev in [95], referring to prior work by Barham, proposed the concept of an 'epistemon' as a description of the sub-systems present in biological functions. Barham [106] regarded living cells as 'epistemic engines', in which a low energy or regulation (epistemic) stroke and a high energy or work (pragmatic) stroke constitute the work cycle. A similar representation was introduced much earlier by the founder of biosemiotics Jakob von Uexküll [109,110]. In fact, Barham may just have rediscovered the latter's semiotic cycle, both phases of the cycle are connected in such a way that the

---

[9]   Whether Being can be talked of in terms of proper parts is an interesting question which has been discussed by Capurro but will not concern us further here.

low-energy (informational) constraints act as signs with respect to high energy (pragmatic) constraints, leading to semiotic, epistemic correlations that have predictive value and which can insure a semiotic, epistemic correlation between the measurement of low-energy environmental signals and the response. The recognition, through its active site of an enzyme to an external chemical stimulus, based on its structure, is such an epistemon. This reaction cannot, at the present level of knowledge, be predicted from the structure *a priori*, but the two components in interaction can be said to be joined by a relation possessing semiotic character. The stability and reproducibility of this relation is made possible via operation of a second semiotic subsystem—the encoding (digital) system. Biological systems therefore include two semiotic subsystems, one based on the structure of imprint and on the recognition of three-dimensional shapes (images), and the other based on the digital linear structure of code.

The interpretation currently preferred by both Authors is based on living systems being first and foremost real, physical, dynamic systems, with a corresponding ontology and hence describable by the 'onto-logic' of Logic in Reality. As in the previous Section, the semiotic descriptions of the meaning-laden biological processes are not false, but are abstractions from the actual flows of energy and molecules in progress. To try to capture these notions, we replace the concept of an 'epistemon' by that of an 'ontolon', an identifiable but dynamic 'individual' (cf. Krause [37]) that refers to or better implies a portion, recognized as such, of that ontological process. As such, an ontolon is a quasi-individual, individual and non-individual, and is an analogue of a quantum system, but only an analogue, quantum-*like*. The ascription of *essential* quantum properties to brain processes is an epistemic ascription.

In contrast to the previous picture, however, our current focus is on actual change rather than on the interpretation of that change, rendering the designation of 'semiotic' almost superfluous. We say almost in order not to exclude epistemological interpretations, but only to insure the proper ontological priority to the dynamics and its logic—LIR. Semiosis in LIR thus describes a *process*, not the application of a doctrine. As a natural process, it is something that starts and stops, and is more or less complete or correct, even persuasive or unpersuasive.

As hypothetical entity, however, the ontolon as some kind of *unit* of real processes-in-progress suggests a kind of identity. We retain for discussion the possibility of a similarity between an ontolon and the Epicurean 'atom' possessing the property of a *clinamen* described above. It thus appears to conflict with our general philosophy of avoiding epistemic identities to describe complex, dynamic phenomena. We argue that there is no *a priori* reason and not to see a process as composed of parts, provided it is understood that these 'parts' are not absolutely separated from a 'whole'. Lupasco showed that a new mereology was implied by the Principle of Dynamic Opposition mentioned above.

Ontolons are thus real, but unlike the epistemological counters of Peirce, we do not require any superaddition of classification or categorization. One can use the concept of ontolons as a device to help focus on the ontological characteristics of processes in relation to other critical aspects such as their ethical valence.

To complete this preliminary discussion of the ontolon as a concept of the parts or entities of dynamic processes, let us briefly consider three other attempts to identify or characterize them. Capurro [111] sees the clear historical derivation from the Greek *ta onta*, a neuter plural participle form best translated by 'existings' or beings. The singular, *to on*, a being, understood as a verbal, participle hence 'dynamic' form would be the equivalent of the ontolon. This interpretation is only acceptable to us if being here includes *becoming*, the processual activity of reality. In our opinion, any attempt to totally separate being from becoming is incorrect. Support (not proof; we are not in a domain of 'proof') comes from the perhaps little-known fact that in several Indo-European languages, the original word for *be* and *become* as the same.

6.2. Ontolons and Information. The Informosome

Following the statement that meaningful information is reality in potential form, we will discuss here how the action of information takes place as a dynamical semiotic process in which the contradictions co-existing as potentialities are resolved during their actualizations and form the basis of new contradictions. We will show that before studying the dynamic process in physics, biology or sociology, we should define the basic ontolon structure for each system. In the frames of this structure, the contradictions become fixed and the system is able to perform a transition to a new level of complexity realizing the principle of negation of negation. By identifying the dynamical structure of ontolon, we understand the basis of a dialectical discourse in the course of interaction between ontolons. A continuous reproduction of a semiotic relation in the structure of ontolon occurs via the process representing the transfer of information.

Peirce coined the neologism 'phaneron' to designate something like the *idea* of Locke forming the 'immediate content of awareness' as opposed to phenomenon. "The phaneron is what appears, as it appears, even if, as it appears, it appears to be more than mere appearance". The difficulty is that Peirce made no ontological commitment regarding his concepts. He wrote specifically that his 'phaneroscopy' (phenomenology) had nothing at all to do with the question of how far the 'phanerons' it studied correspond to any realities. Elsewhere Peirce says that the phaneron is the totality of what is "in any sense present to the mind, quite regardless of whether it corresponds to any real thing or not." We can only agree with Short [112] that Peirce's positions were "fragmentary, marvelously inconsistent and often unpersuasive". We will attempt to tie our ontolon as closely as possible to the evolution of real physical, biological and cognitive processes.

First, the distinction between *res extensa* and *res cogitans* as the basic attributes of the primary substance in the sense of Spinoza defines the two aspects of movement. While the mechanical cause of movement is well assimilated by physics and can be referred to Aristotelian material and formal causes, the cause associated with *res cogitans* is of a different nature and can be associated with Aristotelian efficient and final causes. It is not necessarily associated with consciousness; rather it is a prerequisite of consciousness at higher levels of complexity appearing during the self-movement of the substance that includes both mechanical and non-mechanical constituents. The basic function of cognition, according to Spinoza, is to move and arrange external objects in which mechanical forces can be controlled but not directly involved.

As discussed in Section 2, in Epicurean terms, the non-mechanical cause will correspond in the control of *clinamen* by the group of atoms that use their own *clinamina* to restrict *clinamina* of other atoms in the organized system. In other words, which are relevant to modern quantum mechanical terms, a group of these atoms appears as a device (representation of an "agency") that measures potential trajectories of other atoms and imposes control over their movement, i.e., it navigates them. The unpredictable deviation of these atoms from the established trajectory becomes a prerequisite of their incorporation into a complex system where their potential deviation is restricted. The complex system of atoms acquires the non-mechanical property of internal measurement [113] in which it is subdivided into levels and attains control. The controlling part (navigator) and the controlled (navigated) part are interrelated within the cyclic structure possessing the property of retrocausality [114] in which the transfer of information takes place and the potential state is taken as the meaningful prerequisite of continuous realization of the stable self-supported state of the system. The action that appears as voluntary in fact represents an example of non-mechanical causality that at higher levels of organization becomes mental causality. When the measuring device represents a part of measured system, the measurement proceeds internally in relation to the whole system. The system undergoes complexification in the course of such continuous internal measurement because it leads to an iterative recursive process. This process corresponds to evolution of the whole system.

The internal measuring device measures, among other elements of the measured system, itself in the state of measuring, which limits possible accuracy of quantum measurement and manifests itself as the uncertainty principle [115]. The increase in complexity as a result of quantum measurement corresponds to the increase in complexity in the course of evolution of the system which can be viewed as a potentially infinite recursive embedding process. There was an attempt to explain the visible expansion of the Universe as a consequence of this infinite recursion [116].

The concept of a dynamic 'unit' of process experience can be found in Wu's concept of an 'informosome'. In his philosophical classification Wu Kun [117,118], any existing material structure contains in itself its 'condensed' history, its current properties and the information of its possible or potential future development. It is a 'condensation' of the contents of all relevant operative relationships. This term is currently in use in biology [119] to refer to mechanisms of protein transfer in the cell, but this process should indeed be understood as informational in the broad sense of this paper. This is a further consequence of our view that both material processes and their informational components evolve together [120]. The term exposome refers to the totality of environmental exposures of an individual from conception onwards, and has been proposed as a critical entity for disease etiology. We note that, interestingly, that like the informosome, the exposome is constituted by a totality of *information*. It *is* an informosome.

6.3. Ontolons and Semiosis in Living Systems. Second-order Non-linearity and Poincaré Oscillators

When we consider a cyclic structure with internal semiosis, we may imagine such structures in living systems; however, even in simple organized physical systems such as the hydrogen atom, the controlled restriction in the field of potentialities takes place. An electron in its conjunction with proton can be actualized within certain restricted spatio-temporal area, and the spontaneous collapse of wave function becomes restricted within the organized structure of the atom. This reduction of the field of potentialities bears the information that elementary particle acquires within the atom. At the next level of interaction, the cyclic dynamic structure can recognize other structures that are external to it and react in a way that aims to resist possible destruction that can appear as a result of direct mechanical contact.

Nonlinearity and non-linear dynamics have become substantial new subjects of research in their own right, with two journals being published with those titles! The simplest description of a non-linear system is one in which the change of the output is not proportional to the change of the input. Generally, one tries to use a non-linear system of dynamical equations to describe a complex system. A complex system is loosely defined as constructed by a large number of simple, mutually interacting parts, capable of exchanging stimuli with its environment and of adapting its internal structure as a consequence of such interaction. The non-linear interactions involved can give rise to coherent, emergent complex behavior with a rich structure. Key concepts in complexity science are, for example, the coexistence of diversity and stability, for which LIR provides an interpretation. Complexity science also looks at the dynamics of systems in transition regions of self-organized criticality, but as these are difficult to solve, they are approximated by linear ones, but some phenomena may be hidden by this strategy such as chaos, singularities, and solitons or, we now would add, ontolons. Thus the dynamic behavior of a non-linear system may appear to be counterintuitive and unpredictable.

For us, this picture is quite primitive, since the interactions at the heart of a process and their movement from actuality to potentiality is obscured. We see the term non-linear as describing the complex path between prior and subsequent states of a process from both mathematical and physical perspectives and, we now add, logical ones as well based on Logic in Reality. Thus, *any* movement from cause to effect can be seen as a highly non-linear, multi-dimensional process.

But in what exactly does the non-linearity of non-linear dynamics consist? Many physical phenomena are described as emergent: tornadoes certainly arise from complex temperature and humidity gradients, and such systems are considered to be the consequence of non-linear dynamic interactions. From the LIR standpoint, they are (almost) pure, actualized macrophysical processes with no form of internal representation or semantics. Other examples are the dissipative, far-from-equilibrium systems described by Prigogine, and the intrinsically simple structures such as the convection cells in heated liquids or certain oscillating chemical systems, as discussed below.

Our conclusion from this brief overview of the standard notion of non-linearity is that it is *not* a domain to which the principles of LIR apply. We might call the phenomena referred to, to try to characterize the relation, as first-order non-linear processes. If we look at phenomena instantiating the properties of recursion and self-reference, and, consequently, of the Principle of Dynamic Opposition, we start to see an underlying unity which justifies referring to them as instantiating second-order non-linearity. Further parallels can be drawn to concepts of Synchronic Downward Causation.

We propose that this semiotic system first described by Jacob von Uexküll in relation to living systems [109] represents the basic structure of an ontolon as the element of actualized substance bearing not only spatiotemporal but also cognitive attributes. In a different context, this structure was introduced by James Barham [107] in his model of the dynamical informational semiosis. In his theory of the meaning of information he identifies meaningful functions with generalized non-linear oscillators and their associated phase space attractors. By postulating the existence within such oscillators of a component capable of coordinating low-energy interactions with the correct environmental conditions supporting the dynamical stability of the oscillator the meaning of information is interpreted as the prediction of successful functional action. This can be considered as elementary structure of an ontolon, extending Barham's definition of 'epistemon'. The epistemic function should be considered as a secondary to the ontic nature of such structure that represents the basic unit of nature that can potentially evolve into a complex cognitive system.

In fact, the original invention of such structures should be attributed to Henri Poincaré who introduced non-linear dynamics and differential topology in physics in his concept of the Poincaré map describing the simplest non-linear system called the Duffing oscillator. Glass and Mackey [120] suggested that all systems exhibiting limit-cycle behavior should be termed Poincaré oscillators. This structure can be viewed as the basic structure of the ontolon that is realized at different levels of organization. The periodic orbit of the continuous dynamical system is stable if and only if the fixed point of the discrete dynamical system is stable. The trajectory of such oscillators can be called a periodical attractor or limit cycle. The simplest Poincaré oscillator does not assume a developed internal structure, but a complex internal structure can potentially arise as a result of its evolution towards the expansion of its external stability supported by more sophisticated internal dynamics attributed to its "agency". What is important to point out here is that the interaction of Poincaré oscillators having internal information dynamics can be described as a discourse, which result may not be deterministically anticipated. The oscillator itself has its internal rule that appears as its internal logic, and for its revelation in our description as a meta-logic, it is necessary to describe its behavior as a dialectical process.

While classical mechanical movement can be described by bivalent formal logic, the description of the movement that involves control of potential realizations involves another logical scheme that incorporates the potential-actual states as prerequisites of transition (LIR). This meta-logic is the logic of non-mechanical causality and it can be defined as a dialectical logic in its broad sense. It treats causality as the achievement of a solution established in the discourse of communicating ontolons. In this communication, ontolons can oppose each other or unite in larger ontolons in which the unified system of control is formed. We will further consider the internal structure of ontolons corresponding to living systems and to conscious systems. But if we consider all hierarchies of ontolons, we should briefly return to non-living reality and emphasize that even at the lowest quantum level the reality is can be seen with two attributes, one spatial and the other dynamic. This is reflected in the characteristics of quantum as a "wave-pilot" in the de Broglie interpretation [121].

In relation to the phenomenon of life, the ontolon structure is based on the internal constraints that support the uniqueness of each biological system and its self-maintenance. In this representation, it is possible to understand how biological systems enter into a dialectical discourse that corresponds to the operation of the LIR Principle of Dynamic Opposition. It is thus an oversimplification to view biosystems as simply separable into two functional parts, corresponding to a hardware and software. This static approach is still popular in biosemiotics (see previous Section 5) and other rather reductionist concepts. Such concepts ignore the dynamic aspect of semiosis which consists in the fact that the significant system of the genome is internally reproduced within the system and is repaired through sets of internal constraints representing the feedback from the elements that it encodes. Such a system represents a whole with a locally stable point attractor of the Poincaréan type.

The apparent independence of symbol systems from physical laws follows from the analysis of Poincaréan type oscillators representing the basis of the semiotic dynamical cycles described by von Uexküll and Barham. The view that genetic symbol systems have evolved so far from the origin of life and that semiotics does not appear to have any necessary relation whatsoever to physical laws is also true, but it occults the fact that the processes involved instantiate the same categories of Dynamic Opposition and Non-Separability. Pattee [122] states that information does not belong to the category of universal and inexorable physical laws but refers to initial conditions and boundary conditions. Boundary conditions formed by local structures are called constraints, while informational structures such as symbol vehicles are a special type of constraint. Pattee [123] claims that life and evolvability require the complex interaction of rate-independent symbol constraints and rate-dependent physical dynamics. However, the concepts of initial conditions and constraints in physics make sense only in the context of the law-based physical dynamics to which they apply. This is also the case for the concept of information. We agree with Pattee that the illusion of isolation of symbols from matter can arise from the arbitrariness of the apparent epistemic cut. Further, the apparent isolation of symbolic expression from physics seems due to an epistemic necessity, but ontologically it is still an illusion; making a clear distinction is not the same as isolation from all relations. In general, one clearly separates the genotype from the phenotype, but from an operative standpoint, one certainly does not think of them as isolated or independent entities.

Further elaboration of the matter-symbol problem is possible using the two-level framework of analysis implied by LIR. If the illusion of isolation is an epistemic illusion, whose reality is accepted, this means that symbolic expression is not metaphysically isolated from physics. Consequently, their relation or interaction is real, and it can be considered to have an appropriate dynamics. The remaining question concerns the use of antagonism or constraints to characterize these dynamics. This can be resolved by a view of symbolic memory constraints as dynamic processes in themselves, co-evolving with the other components of biological systems [1].

6.4. Rosen's Model and an Endoperspective

Besides the error-correcting cybernetic controls, the system contains anticipatory mechanisms to pre-empt possible errors, as discussed by Rosen and others [124]. These controls are realized through the agency of a predictive model, converting present information into predicted future consequences. These are *ipso facto* essentially potential and their evolution follows the rules of LIR.

Rosen suggested an alternative to a classical dualistic genetic model of the biological system. He called it the (M, R) system where M is metabolism and R is repair or, as later suggested, replacement [125]. Elements of the metabolic system are continuously replaced and the elements that replace them are also replaced, and this can go to the infinite regression. However, Rosen stated that the system can be "closed to efficient causation" and contain the internal principle of organizational invariance [125] which results in avoiding infinite regression and closing the system in a stable non-equilibrium state in which the system, remaining open to the material flows, becomes selective for them and closed to the efficient causes that are locked inside of it. This structure is based on the principle of "organizational invariance" [126] that escapes the infinite regression for the internal system's time T, during which the system stably performs its function and avoids "global system failure".

Rosen's theory formulates the basic ontolon structure for living systems. Other approaches include Eigen's theory of hypercycles [127] and the autopoietic theory of Maturana and Varela [128] which have common features to Rosen's model but remain less developed in relation to their formalizable and hence logical and ultimately computational value, where possible. Rosen's (M, R) system includes the operation defined by β which designates the system as a whole and acts as a generator of the complete enclosed structure of (M, R) systems. There is no rigid algorithm to take β as this operation has its own ambiguity, but when the β is taken, this ambiguity becomes frozen and internal computation becomes possible (cf. the 'frozen dialectics' of Lupasco [35]. Later, Gunji et al. [129] came to the idea of describing the organizational invariance of a biological system from the point of view of "heterarchy" which naturally involves self-reference through the inherited logical inconsistencies between levels. One description of a heterarchy is as a dualism of the property of self-reference and the property of a frame problem, interacting energetically with each other. The research program of Gunji represents an important current development of Rosen's ideas in which a successful explanation of internally constrained principles of animal behavior and human consciousness has been reached [129]. Rosen's understanding of organizational invariance is similar to Gödel's encoding of statements about the system within it in the application to biological systems. The β parameter representing organizational invariance is equivalent to an agent establishing the set of Gödel numbers generated within the system. The whole biological system can be viewed as consisting of: (a) Metabolism—sets and relations; (b) Replacement—relations on relations; and (c) Organizational invariance—Gödel statements about the sets actually in place and their dynamic relations. It is important to note that Gödel statements are not sets or relations; they are meta-mathematical statements with the dual function of both sets and relations. Logic in Reality is the first *metalogical* system in which such statements can be embedded and given a real interpretation. Thus, the triadic structure of life includes sets, relations, and meta-mathematical as well as metalogical statements (encoded within the system) that are as real and as causally efficient as much as any physical element of the system. Translation of these statements into the system occurs via a sophisticated 'machinery' of transcription, translation and recombination, e.g., splicing events. But we remind the reader that any such event is a consequence of the real dualities or polarities in chemical and biological systems and the movement within them between potentiality and actuality and vice versa.

6.5. The Ontolon Structures of Social Consciousness and Systems

When we turn to individual and social systems instantiating reflexive consciousness, we find more advanced ontolon structures of the Poincaréan type. The theory of functional systems of Pyotr Anokhin [130] explores the basic ontolon structure for such systems. These systems are based not on the linear transmission of information from receptors to executive organs but composed of synchronized distributed elements organized non-linearly. They support homeostasis due to a change of behavior occurring in the course of interaction with the outside world. The Poincaréan systems including Uexkull's biosemiotics system and, in relation to cognitive events, Anokhin's functional system represent these basic non-linear structures viewed externally, partly but not totally separated from the internal reflexive relation viewed from an endoperspective. The endoperspective in the dynamics of organized systems was explored earlier in dialectical discourse starting with Heraclitus, Plato and Aristotle. The movement of science towards an endoperspective today represents a new level of understanding that as demonstrated by the concepts of Rosen and Gunji, to which the principles and dialectics of Logic in Reality apply.

Social communication, in the light of the LIR approach to communication outlined in Section 5.6, becomes a prerequisite of reflexive consciousness. According to Lev Vygotsky and further to Evald Ilyenkov, all higher psychic functions are the internalized social relations. They determine first the interdictive relations where certain actions are prohibited and then turn to the suggestive phase where these prohibitions are internalized (Boris Porshnev [131]) and the person becomes the other for himself and by this controls, regulates and changes his own activity towards other individuals and the external world in total. Almost identical language for the internalization of the nominally external 'other' was used by Igamberdiev to show the ultimately logical rationale for ethical behavior [60]. The work studied by Brenner [1] provides a more detailed scientific explication of Lupasco's insights in this area.

The prerequisite of domination is another important feature of nervous system called the law of dominance formulated by Alexei Ukhtomsky. By "dominant" Ukhtomsky defined "a more or less stable focus of increased excitability, evoked in whatever way, and stimuli newly arriving at the centers of excitation serve to amplify (reinforce) excitation in this focus, while in the rest of the central nervous system inhibition spreads widely" [132]. Through the dominance of the reflexive excitation in nervous system, the "internal speech" is formed that can be represented as a reflexive relation of the structure of the subject. The mechanism of reference of the subject to himself represents an elementary unit of reflexive consciousness. It also determines the relation of a human being to other human beings which basis is the system of mirror neurons that fire both when the subject acts and when he observes the same action performed by another. The system of references to him/herself and to others leads to the formation of discrete reflexive types that contradict to each other, interact and form complex social relations. This was developed further in the reflexive psychology of Vladimir Lefebvre.

In psychology, the subject becomes a part of a system where it can reflect himself. In this system, the biological reality is represented as unconscious and interpreted as *Ego* through signification by the image of other, *Superego* (Freud) or *Symbolic* (Lacan). It can be considered another logical pattern describing interrelations between consciousness and the external world. It determines the fixation of somebody's image into the other as a possibility to substitute the other [133]. In formalization of this model by Lefebvre [134], the subject A_1 constructs the image of himself (a_2) and his image's image of himself (a_3). The subject's state a_1 will be a composition of the contradictory statements a_2 and a_3. Thus, the subject will correspond to a character $A_1 \equiv a_1a_2a_3$. The labor activity of humans is accompanied by the semiotic internalization of tools in the language. The thesis of Vygotsky [135] about the internalization of labor tools in the signs of language is incorporated into this paradigm. To link the labor activity to mental development, the internalization factor is necessarily similar to the "organizational invariance" β in Rosen's theory. This transition reflects the ideas of Vygotsky on development as a process of internalization. The bodies of knowledge and tools of thought evolve and are linked together to form a new dynamic entity of the human culture—an ontolon—internalizing the environment. Development consists of the gradual internalization of the environment, primarily through the language, to form cultural adaptation. Piaget [136] considered that several stages of such internalization during the development of the individual child reflect an evolutionary succession. In other words, such internalization proceeds via the establishment of a relation to the external reality.

6.6. Experimental Metaphysics. Categories

These considerations provide an occasion to say what our approach, in the 'language' of ontolons and epistemons, is *not*. It is not an 'experimental metaphysics' in the concept of Shimony and of Redhead [137]. Methods of valid argument in current philosophy still embody the tautological assumptions of classical, propositional logic and its notion of truth. Redhead's (well-intentioned) movement "from physics to metaphysics» moved further away, not towards, reality.

The metaphysical world-view that is implied by the Principle of Dynamic Opposition (PDO) is compatible with the 'metaphysical revision that has been engendered by quantum mechanics', in the phrase of Vlatko Vedral [138]. One does not have to have a prior 'orthodox' concept of reality in order to define the best possible active role for what is observed, namely, that dualities are present at all levels of reality, starting with that of the quantum field. The dualities in question have a kind of part-whole relation to the world, but one need not assume that at the end of this analysis, one will have captured all the essential aspects of the world. As noted above, one will not have, as a consequence, a 'Theory of Everything' (which was not an objective in the first place), but one will have a framework that can evolve in parallel with further development in the physical understanding of our universe. Our view is consistent with the work of Vedral.

Logic in Reality, as outlined in [1] includes a 'New Energy Ontology' which includes categories but redefines their characteristics. The absolute, binary concepts, exclusivity and exhaustivity are eliminated in favor of a dynamic relation between 'objects and forms' of thought. The role of such categories in ontology is essential in defining LIR as a conceptual structure that has additional explanatory power. In a categorial realist conception, as suggested by Thomasson [139], "providing a system of categories can be seen as a, or even *the* central task of metaphysics". We believe a robustly realist position is made more plausible by the principles of LIR, since they improve our ability to discern intrinsic divisions and redefine changes or movements in physical reality as the "entities" in the category of existence. For our purposes it is not necessary to decide for an ontological or metaphysical reading of the term 'category': both can be used as they complement one another.

6.7. The Ontological Priority of Ontolons

In contrast to the concept of 'monads' (cf. Leibniz) the notion of ontolons as entities or forms of existence—'being in reality'—has received far less discussion. We propose ontolon as our basic term for the essences of beings-in-reality, for the *forms of existence* which assume a multiplicity of dynamic processes in the extant domain, including knowing. The epistemon, then, represents the internal semiotic structure of ontolons, their epistemic representation. Logic in Reality provides the framework for the description of both ontic and epistemic processes and their relations, united via the transcending cognitive logical operations involving potential and actual states in interaction. Taken to its logical (*sic*) conclusion, the notion of ontolon is better grounded in reality than that of monad. It can replace the term of monad and, eventually, that of the *Dasein* of Heidegger in a Philosophy in Reality in which the number of non-natural concepts are reduced to a minimum.

7. Summary and Conclusions

In the spirit of this paper, and its emphasis on process rather than products, we see our 'conclusion' also as not final but as a pointer toward processes leading to further and better descriptions of reality. Our description here of reality and its philosophy simply assumes that we apply certain schemes of valid reasoning which are nevertheless based on reasoning as an ontological as well as epistemological process. The reasoning can still be formal or informal. Formal reasoning has proven to be a powerful tool for understanding our world. However, the application of formal reasoning, like any human cognitive process, is partly informal and cannot be fully grasped in terms of formal logic alone. This process is *dialectical* in that if different solutions are present at the same time the optimal one is chosen. It has been described for example by Magnani [140], following Peirce, as abduction, but in our view this approach is limited to an epistemology. It does resemble the search for proper Gödel numbers to analyze the validity of formal description in the proof of the incompleteness theorem. Wittgenstein was the first who suggested that such epistemological processes appear as language games, and their formalization can be processed as description of moves in games. This is achieved in the theory of defeasible reasoning, which is the reasoning that is rationally compelling, though not deductively valid. The approaches to set the limits of formalization of dialectics have been developed by such philosophers as Nicholas Rescher [141], Frans H. van Eemeren and Rob Grootendorst [142], but, again, mainly in the relation to argumentation in communication as solely a linguistic process. A more advanced dynamic approach with application to the nature of consciousness itself is being developed by Yukio-Pegio Gunji et al. [143]. They propose a measurement-oriented inference system comprising Bayesian and inverse Bayesian inferences. In this model, Bayesian inference contracts the probability space while the inverse inference relaxes it which allows a subject to make a decision corresponding to an immediate change in the environment. To date, this model with two inferences represents the best attempt at a representation of the dialectical discourse by means of formal models.

To the extent that dialectics also refers to informal reasoning, the question arises how discourse based on informal reasoning is possible in the real world before the appearance of consciousness. As Engels [15] stated, "nature is the proof of dialectics", and such statements sound like the action of Diogenes who proved the reality of movement by walking as his 'dynamic' refutation of Zeno's paradoxes. In fact, dialectics is an example of informal reasoning itself. The answer to our question lies in the fact that the Being that becomes the Being-in-Reality (*Dasein* in the Heideggerian terms) appears as a multiplicity of the forms of existence for which we have proposed the term ontolons. (Our 'multiplicity of forms' is clearly related to the insights of classical Chinese formulations [144]).

Ontolons are thus the dynamic counterparts of well-known prior entities, namely, the monad and the 'Thing-in-Itself'. Each ontolon, besides the property of deterministic mechanical movement arising from its spatio-temporality (*res extensa*), possesses the property that manifests the internal self (*res cogitans*). Even in its simplest forms it is exhibited in unpredictable (non-deterministic in mechanical sense) movement. This is what Epicurus called the *clinamen*. Being unpredictable originally, it can become controlled in communication between ontolons, in which a reduction of potentialities for totally free movement takes place.

In reality, of course, we are now in the domain of the further *res potentia* of Heisenberg, reviewed recently by Brenner in [145]. In this discourse, ontolons can be seen as forming more advanced structures, in which potentiality inheres *a priori*, leading finally to the appearance of epistemic properties corresponding to life and finally consciousness. The reality in this framework can be represented as a set of forms of existence capable of a continuous process of complexification in which the most optimal realizations can occur. This is what Luhn [146] has called, coming from the side of Information Science, the 'search' by the universe for new dynamic states.

Self is seen as the principle that governs *clinamina*. We cannot see the self as we do not have "window" (in the Leibniz sense) to see it but we can logically deduce its existence through the observation of reduction of potentialities that takes place and thus we have access to "Processes-in-Themselves", as ontolons. We become involved in the Processes-in-Themselves in a such a way—better we *are* those processes—that we are able to apply formal reasoning to them (and eventually compute those which are computable.) But, to repeat, the process of discovery via formal reasoning remains informal and appears as a dialectical process occurring in nature. A very suggestive link can be made here to the concept of Natural Computation currently being developed by Gordana Dodig-Crnkovic [147] and others. As stated above, it should be described within a framework that includes both ontic and epistemic processes and their relations, united via the transcending cognitive logical operations involving potential and actual states in interaction. The logic of the included middle introduced by Lupasco [35] in its current version as Logic in Reality [1] provides such a framework.

Principles and the Common Good

This paper is about principles of science, philosophy and logic, but it is also a statement *of* principle: we do not believe that any philosophic 'work' of the kind we are engaged in can be justified, as can pure science, if it is 'pure' philosophy. The use of human and natural resources can be justified for science, as such and as a part of natural philosophy, *because* is directed at increases in understanding of the real world. At the other extreme, we have the discussions, following Kripke and others, of possible worlds that have no existence other than that of fictional objects.

There is another operative principle that we can state as a conclusion of this study: any synthesis we seek will not be dependent on any absolute criteria of truth or completeness, but will seek to incorporate or in any event refer as far as possible to contradictory or opposing points of view. The second is that, as a metaprinciple, one can talk about principles at all, which some have contested. We are left with a criterion of utility like that of J. S. Mill which has been considered 'weak' and non-scientific. But this is exactly our thesis, even though by our own criteria we cannot prove or justify it, but support it on a methodological basis, that is, that dialectics and the Lupascian logic give the least abstract possible picture of the world.

Our dialectical approach leads to recovery of both dialectics and semiotics from reductionist interpretations and to their reunification in a new synthetic paradigm centered on meaning and its communication. Our concept unites science, logic and philosophy in a common meta-thesis and provides the real contours of a basic understanding of nature and civilization.

Formalization of dialectical logic cannot be complete and that is why it tends, like other 'diversities', to be ignored in scientific discourse [29]. However, the logic of real processes to redefine the ontological relations between meaning, communication and language will always remain a fundamental task; it forms the background of any description of nature that can accompany the new functional convergence of science and philosophy in progress. We consider the development of this 'dialectical realism' as a basis of the ethical development of knowledge for the common good.

Author Contributions: Conceptualization, J.E.B. and A.U.I.; Writing—original draft, J.E.B. and A.U.I.

Funding: This research received no external funding.

Acknowledgments: The authors acknowledge with gratitude the friendly and critical support of Rafael Capurro, Wolfgang Hofkirchner and Wu Kun. We look forward to an on-going dialogue with them on the philosophical and scientific issues addressed in this paper.

Conflicts of Interest: The authors declare no conflict of interest.

References

1. Brenner, J. *Logic in Reality*; Springer: Dordrecht, The Netherlands, 2008.
2. Wu, K. The Interaction and Convergence of the Philosophy and Science of Information. In Proceedings of the 2nd International Conference on the Philosophy of Information, Vienna, Austria, 6 June 2015.
3. Wu, K.; Brenner, J. The Informational Stance: Philosophy and Logic. Part I: The Basic Theories. *Log. Log. Philos.* **2013**, *22*, 1–41.
4. Boole, G. *The Laws of Thought. On Which Are Founded the Mathematical Theories of Logic and Probabilities*; Dover Publications, Inc.: Mineola, NY, USA, 1958; p. 424, Originally Published by Macmillan, NY, 1854.
5. Deleuze, G.; Krauss, R. Plato and the Simulacrum. *October* **1983**, *27*, 45–56. [CrossRef]
6. Alcoff, L.; Shomali, A. Adorno's Dialectical Realism. *Symp. Can. J. Cont. Philos./Rev. Can. De Philos. Cont.* **2010**, *14*, 45–65. [CrossRef]
7. McGinn, C. *Basic Structures of Reality. Essays in Meta-Physics*; Oxford University Press: Oxford, UK, 2011; p. 174.
8. Dodig-Crnkovic, G.; Schroeder, M. Contemporary Natural Philosophy and Philosophies. *Philosophies* **2018**, *3*, 42. [CrossRef]
9. Brenner, J. The Naturalization of Natural Philosophy. *Philosophies* **2018**, *3*, 41. [CrossRef]
10. Igamberdiev, A.U. Time and life in the relational universe: Prolegomena to an integral paradigm of natural philosophy. *Philosophies* **2018**, *3*, 30. [CrossRef]
11. Stent, G.S. *Molecular Genetics; an Introductory Narrative*; W.H. Freeman: San Francisco, CA, USA, 1971.
12. Aristotle. On the Soul. In *The Complete Works of Aristotle*; Barnes, J., Ed.; Smith, J.A., Translator; Princeton University Press: Princeton, NJ, USA, 1995; pp. 641–692.
13. Rovelli, C. *The First Scientist. Anaximander and His Legacy*; Westholme Publishing: Yardley, PA, USA, 2011.
14. Kant, I. *Critique of Pure Reason*; Cambridge University Press: Cambridge, UK, 1998; Originally published 1781.
15. Engels, F. *Dialectics of Nature*; Progress Publishers: Moscow, Russia, 1976; Originally published 1877.
16. Needham, J. *Moulds of Understanding*; George Allen & Unwin: Crows Nest, Australia, 1976.
17. Mamardashvili, M.K. *Cartesian Meditations*; Progress: Moscow, Russia, 1993. (In Russian)
18. Ilyenkov, E.V. *Dialectical Logic, Essays on its History and Theory*; Aakar Books: Delhi, India, 2008; Originally published in Russian 1977.
19. Voevodsky, V. A1-homotopy theory. *Proc. Int. Cong. Math.* **1998**, *1*, 579–604.
20. Hume, D. *Dialogues Concerning Natural Religion*; Cambridge University Press: Cambrdige, UK, 2007; Originally published 1779.
21. Smolin, L. *The Life of the Cosmos*; Oxford University Press: Oxford, UK, 1997.
22. Brenner, J. Consciousness as Process: A New Logical Perspective. *APA Newsl.* **2018**, *18*, 10–22.
23. Brenner, J. The Philosophy of Ecology and Sustainability: New Logical and Informational Dimensions. *Philosophies* **2018**, *3*, 16. [CrossRef]
24. Nicolescu, B. *Nous, la particule et le monde*; Editions du Rocher: Paris, France, 2002; (Originally published in Paris: Éditions Le Mail, 1985).
25. Information and Communications Technologies and Society. In *Proceedings of the Summit Conference on Information*; International Society for Studies of Information: Vienna, Austria, 2015.
26. Brenner, J. Three Aspects of Information Science in Reality: Symmetry, Semiotics and Society. *Information* **2015**, *6*, 750–772. [CrossRef]
27. Wu, K.; Brenner, J. Informational Thinking and Systems Thinking: A Comparison. In Proceedings of the Foundations of Information Conference, Moscow, Russia, 13 June 2013.
28. Hofkirchner, W. *Emergent Information. A Unified Theory of Information Framework: World Scientific Series in Information Studies*; World Scientific: Singapore, 2013; Volume 3, p. 239.
29. Brenner, J. Process in Reality: A Logical Offering. *Log. Log. Philos.* **2005**, *14*, 165–202. [CrossRef]
30. Brenner, J. Information in Reality: Logic and Metaphysics. *Triple-C* **2011**, *9*, 332–341. [CrossRef]

31. Roy, S. Quantum Probability Theory and non-Boolean Logic. In *Decision Making and Modeling in Computer Science*; Springer: Delhi, India, 2016.
32. Aerts, D.; D'Hooghe, B. Potentiality States: Quantum versus Classical Emergence. *arXiv* **2012**, arXiv:1212.0104v1.
33. Conte, E.; Khrennikov, A.; Zbilut, J.P. The Transition from Ontic Potentiality to Actualization of States in a Quantum Mechanical Approach to Reality. *arXiv* **2006**, arXiv:quant-ph/0607196.
34. Lupasco, S. *L'univers psychique*; Denoël/Gonthier: Paris, France, 1979.
35. Lupasco, S. *Le principe d'antagonisme et la logique de l'énergie*; Editions du Rocher: Paris, France, 1987; (Originally published in Paris: Éditions Hermann, 1951).
36. Aerts, D.; D'Hondt, E.; Gabora, L. Why the Logical Disjunction in Quantum Logic is not Classical. *Found. Phys.* **2000**, *30*, 1473–1480. [CrossRef]
37. Krause, D. Separable Non-Individuals. *Representaciones* **2005**, *1*, 21–36.
38. Stcherbatsky, F.T. *Buddhist Logic*; Dover: New York, NY, USA, 1962; Originally published in Leningrad: Academy of Sciences of the U.S.S.R., 1930, p. 17.
39. Cao, T.Y. *Conceptual Developments of 20th Century Field Theories*; Cambridge University Press: Cambridge, UK, 1997.
40. Lupasco, S. *Logique et Contradiction*; Presses Universitaires de France: Paris, France, 1947.
41. Taylor, C. *Hegel*; Cambridge University Press: Cambridge, UK, 1975.
42. Sanchez-Palencia, E. *Promenade Dialectique dans les Sciences*; Hermann Éditeurs: Paris, France, 2012.
43. Priest, G. *Contradiction*; Martinus Nijhoff Publishers: Dordrecht, The Netherlands; Boston, MA, USA; Lancaster, UK, 1987.
44. Priest, G. Dialectic and Dialetheic. *Sci. Soc.* **1989**, *53*, 388–415.
45. Lupasco, S. *L'énergie et la matière vivante*; Éditions du Rocher: Monaco, 1986; (Originally published in Paris: Julliard, 1962).
46. Bohm, D.; Hiley, B. *The Undivided Universe. An Ontological Interpretation of Quantum Theory*; Routledge: London, UK; New York, NY, USA, 1993.
47. Arthur, C. *The New Dialectic and Marx's Capital*; Akkar Books: Delhi, India, 2013.
48. Deacon, T. *Incomplete Nature: How Mind Evolved from Matter*; W.W. Norton & Co.: New York, NY, USA, 2012.
49. Lakoff, G.; Johnson, M. Philosophy in the Flesh. In *The Embodied Mind and its Challenge to Western Thought*; Basic Books: New York, NY, USA, 1999.
50. Marx, K. *Das Kapital: Kritik der politischen Oekonomie*, 1st ed.; Verlag von Otto Meissner: Hamburg, Germany, 1867.
51. Carrera, J. The Consciousness of the Working Class as Historical Subject. In *Marx's Capital and Hegel's Logic. A Reexamination*; Haymarket Books: Leiden, The Netherlands, 2014.
52. Bellofiore, R. Lost in Translation? Once Again on the Marx-Hegel Connection. In *Marx's Capital and Hegel's Logic. A Reexamination*; Haymarket Books: Leiden, The Netherlands, 2014; pp. 164–188.
53. Uno, K. *The Types of Economic Policies under Capitalism*; Sekine, T., Translator; Haymarket Books: Chicago, IL, USA, 2018.
54. Sekine, T. *The Types of Economic Policies under Capitalism*; Sekine, T., Translator; Haymarket Books: Chicago, IL, USA, 2018; p. 325.
55. Brenner, J. On Representation in Information Theory. *Information* **2011**, *2*, 560–578. [CrossRef]
56. Petrov, P. Mixing Signs and Bones. John Deely's Case for Global Semiosis. *Signs Syst. Stud.* **2013**, *41*, 404–423. [CrossRef]
57. Burgin, M.; Brenner, J. Operators in Nature, Science, Technology and Society: Mathematical, Logical and Philosophical Issues. *Philosophies* **2017**, *2*, 21. [CrossRef]
58. Gödel, K. *The Consistency of the Axiom of Choice and of the Generalized Continuum Hypothesis with the Axioms of Set Theory*; Princeton University Press: Princeton, NJ, USA, 1940.
59. Igamberdiev, A.U. Time rescaling and pattern formation in biological evolution. *Biosystems* **2014**, *123*, 19–26. [CrossRef]
60. Igamberdiev, A.U. Evolutionary transition from biological to social systems via generation of reflexive models of externality. *Prog. Biophys. Mol. Biol.* **2017**, *131*, 336–347. [CrossRef]
61. Rosen, R. *Life Itself: A Comprehensive Inquiry into the Nature, Origin, and Fabrication of Life*; Columbia University Press: New York, NY, USA, 1991.
62. *Friedrich Wilhelm Joseph Schelling's Sämmtliche Werke*; Schelling, K.F.A. (Ed.) I Abtheilung Vols. 1-10, II Abtheilung Vols. 1-4; Cotta: Stuttgart, Germany, 1861.

63. Gare, A. From Kant to Schelling to Process Metaphysics: On the Way to Ecological Civilization. *Cosm. Hist.* **2011**, *7*, 6–69.
64. Whitehead, A.N. *Process and Reality. An Essay in Cosmology. Gifford Lectures Delivered in the University of Edinburgh During the Session 1927–1928*; Macmillan: New York, NY, USA; Cambridge University Press: Cambridge, UK, 1929.
65. Thompson, I. Process Theory and the Concept of Substance. *Generative Science.* 1990. Available online: http://www.generativescience.org/ph-papers/subst5c.html (accessed on 13 May 2019).
66. Bogdanov, A.A. *Essays in Tektology: The General Science of Organization*, 2nd ed.; Intersystems Publications: Seaside, CA, USA, 1984; Originally published 1921.
67. Gare, A. Aleksandr Bogdanov and systems theory. *Democr. Nat.* **2000**, *6*, 341–359. [CrossRef]
68. Zelený, M. Tektology. *Int. J. Gen. Syst.* **1988**, *14*, 331–342. [CrossRef]
69. Igamberdiev, A.U. Semiokinesis – Semiotic autopoiesis of the Universe. *Semiotica* **2001**, *135*, 1–23. [CrossRef]
70. Igamberdiev, A.U. *Physics and Logic of Life*; Nova Science Publishers: Hauppauge, NY, USA, 2012.
71. Igamberdiev, A.U. Semiosis and reflectivity in life and consciousness. *Semiotica* **1999**, *123*, 231–246. [CrossRef]
72. Igamberdiev, A.U.; Shklovskiy-Kordi, N.E. Computational power and generative capacity of genetic systems. *Biosystems* **2016**, *142–143*, 1–8. [CrossRef]
73. Igamberdiev, A.U. Hyper-restorative non-equilibrium state as a driving force of biological morphogenesis. *Biosystems* **2018**, *173*, 104–113. [CrossRef]
74. Igamberdiev, A.U. *Time, Life, and Civilization*; Nova Science Publishers: Hauppauge, NY, USA, 2015.
75. Sawa, K.; Igamberdiev, A.U. The Double Homunculus model of self-reflective systems. *Biosystems* **2016**, *144*, 1–7. [CrossRef]
76. Lupasco, S. *L'énergie et la matière psychique*; Editions le Rocher: Paris, France, 1987; (Originally published in Paris: Julliard, 1974).
77. Capurro, R. Foundations of Information Science. Review and Perspectives. In Proceedings of the International Conference on Conceptions of Library and Information Science, University of Tampere, Tampere, Finland, 26–28 August 1991.
78. Burgin, M. *Theory of Information. Fundamentality, Diversity and Unification*; World Scientific: Singapore, 2010.
79. Brenner, J. Information. A Personal Synthesis. *Information* **2014**, *5*, 134–170. [CrossRef]
80. Brenner, J. Prolegomenon to a logic for the Information Society. *Triple-C* **2009**, *7*, 38–73. [CrossRef]
81. Diaz Nafria, J.; Zimmermann, R.E. Emergence and Evolution of Meaning. *Triple-C* **2013**, *11*, 13–35. [CrossRef]
82. Brier, S. *Cybersemiotics. Why Information is not Enough*; University of Toronto Press: Toronto, ON, Canada, 2008.
83. Kauffman, S.; Logan, R.K.; Este, R.; Goebel, R.; Hobill, D.; Shmulevich, I. Propagating organization: An enquiry. *Biol. Philos.* **2007**, *23*, 27–45. [CrossRef]
84. Lupasco, S. *Qu'est-ce qu'une structure?* Christian Bourgois: Paris, France, 1967.
85. Deacon, T.; Cashman, T. Steps to a Metaphysics of Incompleteness. *Theol. Sci.* **2016**, *14*, 401–429. [CrossRef]
86. Deacon, T. What is missing from theories of Information? In *Information and the Nature of Reality: From Physics to Metaphysics*; Davies, P., Gregersen, N.H., Eds.; Cambridge University Press: Cambridge, UK, 2010.
87. Ulanowicz, R. Towards Quantifying a Wider Reality: Shannon Exonerata. *Information* **2011**, *2*, 624–634. [CrossRef]
88. Rosen, R. Complexity as a system property. *Int. J. Gen. Syst.* **1977**, *3*, 227–232. [CrossRef]
89. Davies, P.C.W. Does quantum mechanics play a non-trivial role in life? *Biosystems* **2004**, *78*, 69–79. [CrossRef]
90. Igamberdiev, A.U. Physical limits of computation and emergence of life. *Biosystems* **2007**, *90*, 340–349. [CrossRef]
91. Igamberdiev, A.U. Biomechanical and coherent phenomena in morphogenetic relaxation processes. *Biosystems* **2012**, *109*, 336–345. [CrossRef]
92. Rapoport, I.A. *Microgenetics*; Nauka: Moscow, Russia, 1965. (In Russian)
93. Igamberdiev, A.U.; Shklovskiy-Kordi, N.E. The quantum basis of spatiotemporality in perception and consciousness. *Prog. Biophys. Mol. Biol.* **2017**, *130*, 15–25. [CrossRef]
94. Nicolescu, B.; Brenner. Personal communication, 2019.
95. Igamberdiev, A.U. Objective patterns in the evolving network of non-equivalent observers. *Biosystems* **2008**, *92*, 122–131. [CrossRef]
96. Wu, K.; Brenner, J. Philosophy of Information: Revolution in Philosophy. Towards an Informational Metaphilosophy of Science. *Philosophies* **2017**, *2*, 22. [CrossRef]

97. Wu, K. The Interaction and Convergence of the Philosophy and Science of Information. *Philosophies* **2016**, *1*, 228. [CrossRef]
98. Wu, K. The Essence, Classification and Quality of the Different Grades of Information. *Information* **2012**, *3*, 403–419. [CrossRef]
99. Wu, K.; Brenner, J. Chapter 8: A Unified Science-Philosophy of Information in the Quest for Transdisciplinarity. In *Information Studies and the Quest for Transdisciplinarity*; World Scientific: Singapore, 2017. [CrossRef]
100. Wolfram, S. A New Kind of Science. Wolfram Media: Champaign, IL, USA, 2002.
101. Capurro, R. Hermeneutics and the Phenomenon of Information. In *Metaphysics, Epistemology and Technology. Research in Philosophy and Technology*; Mitcham, C., Ed.; JAI/Elsevier Inc.: Amsterdam, The Netherlands, 2000; Volume 19, pp. 79–85.
102. Capurro, R. Interpreting the Digital Human. 2008. Available online: http://www.capurro.de/wisconsin.html (accessed on 3 March 2015).
103. Capurro, R.; Takenouchi, T.; Tkach-Kawasaki, L.; Iitaka, T. On the Relevance of Angeletics and Hermeneutics for Information Technology. In *Messages and Messengers. Angeletics as an Approach to the Phenomenon of Communication, ICIE Series*; Capurro, R., Holgate, J., Eds.; Wilhelm Fink: Munich, Germany, 2011.
104. Holgate, J. *Homo loquens* meets *homo informaticus*; exploring the relationship between language and information. In Proceedings of the 3rd International Conference on the Philosophy of Information, Gothenburg, Sweden, 12–14 June 2017.
105. Brenner, J. Levels of Abstraction; Levels of Reality. In *Luciano Floridi's Philosophy of Technology*; Springer Science + Business Media: Dordrecht, The Netherlands, 2012.
106. Barham, J. A Poincaréan Approach to Evolutionary Epistemology. *J. Soc. Biol. Struct.* **1990**, *13*, 193–258. [CrossRef]
107. Barham, J. A dynamical model of the meaning of information. *Biosystems* **1996**, *38*, 235–241. [CrossRef]
108. Capurro, R.; Brenner. Personal communication, 6 January 2019.
109. Von Uexküll, J. *Theoretische Biologie*; Suhrkamp: Frankfurt-am-Main, Germany, 1973; Originally published 1920.
110. Von Uexküll, J. The theory of meaning. *Semiotica* **1982**, *42*, 25–82, Originally published 1940. [CrossRef]
111. Capurro, R.; Brenner. Personal communication, 15 February 2019.
112. Short, T. *Peirce's Theory of Signs*; Cambridge University Press: New York, NY, USA, 2007.
113. Matsuno, K. Quantum and biological computation. *Biosystems* **1995**, *35*, 209–212. [CrossRef]
114. Matsuno, K. From quantum measurement to biology via retrocausality. *Prog. Biophys. Mol. Biol.* **2017**, *131*, 131–140. [CrossRef]
115. Matsuno, K. The uncertainty principle as an evolutionary engine. *Biosystems* **1992**, *27*, 63–76. [CrossRef]
116. Kineman, J.J. Relational self-similar space-time cosmology revisited. In Proceedings of the 54th Meeting of the International Society for System Sciences, Waterloo, ON, Canada, 18–23 July 2010.
117. Wu, K.; Brenner, J. An Informational Ontology and Epistemology of Cognition. *Found. Sci.* **2015**, *20*, 249–279. [CrossRef]
118. Wu, K.; Brenner, J. The Informational Stance. Philosophy and Logic. Part II: From Physics to Society. *Log. Log. Philos.* **2014**, *23*, 81–108.
119. Allaby, M. *A Dictionary of Plant Sciences*; Oxford University Press: Oxford, UK, 1998.
120. Glass, L.; Mackey, M.C. *From Clocks to Chaos: The Rhythms of Life*; Princeton University Press: Princeton, NJ, USA, 1988.
121. De Broglie, L.V.P.R. *Certitudes et incertitudes de la science*; Albin Michel: Paris, France, 1966.
122. Pattee, H. The physics of symbols: Bridging the epistemic cut. *Biosystems* **2001**, *60*, 5–21. [CrossRef]
123. Pattee, H. The physics of autonomous biological information. *Biol. Theory* **2006**, *1*, 224–226. [CrossRef]
124. Rosen, R. *Anticipatory Systems. Philosophical, Mathematical and Methodological Foundations*; Pergamon Press: New York, NY, USA, 1985.
125. Rosen, R. *Fundamentals of Measurement and Representation of Natural Systems*; Elsevier North-Holland: New York, NY, USA, 1978.
126. Letelier, J.-C.; Soto-Andrade, J.; Abarzúa, F.G.; Cornish-Bowden, A.; Cárdenas, M.L. Organizational invariance and metabolic closure: Analysis in terms of (M,R) systems. *J. Theor. Biol.* **2006**, *238*, 949–961. [CrossRef]
127. Eigen, M.; Schuster, P. *The Hypercycle: A Principle of Natural Self-Organization*; Springer: Berlin, Germany, 1979.

128. Varela, F.; Maturana, H.; Uribe, R. Autopoiesis: The organization of living systems, its characterization and a model. *Biosyst. (Curr. Mod. Biol.)* **1974**, *5*, 187–196. [CrossRef]
129. Gunji, Y.-P.; Sasai, K.; Wakisaka, S. Abstract heterarchy: Time/state-scale re-entrant form. *Biosystems* **2008**, *91*, 13–33. [CrossRef]
130. Anokhin, P. *Biology and Neurophysiology of the Conditioned Reflex and Its Role in Adaptive Behavior*; Pergamon: Oxford, UK, 1974.
131. Porshnev, B. *On the Dawn of Human History (Problems of Paleopsychology)*; Mysl: Moscow, USSR, 1974. (In Russian)
132. Ukhtomsky, A. *The Dominant*; Leningrad State University: Saint Petersburg, Russia, 1966. (In Russian)
133. Lacan, J. *Écrits: The First Complete Edition in English*; Fink, B., Translator; W.W. Norton & Co.: New York, NY, USA, 2006.
134. Lefebvre, V. The fundamental structures of human reflection. *J. Soc. Biol. Struct.* **1987**, *10*, 129–175. [CrossRef]
135. Vygotsky, L. Thinking and Speech. In *The Collected Works of L.S. Vygotsky*; Springer: Berlin, Germany, 1987; Originally published 1934.
136. Piaget, J. *The Language and Thought of the Child*; Routledge & Kegan Paul: Abingdon, UK, 1926.
137. Redhead, M. *From Physics to Metaphysics*; Cambridge University Press: Cambridge, UK, 1995.
138. Vedral, V. *Decoding Reality*; Oxford University Press: New York, NY, USA, 2010.
139. Thomasson, A. *Fiction and Metaphysics*; Cambridge University Press: Cambridge, UK, 1999.
140. Magnani, L. *Abductive Cognition. The Epistemological and Eco-Cognitive Dimensions of Hypothetical Reasoning. Cognitive Systems Monographs Vol. 3*; Springer: Berlin/Heidelberg, Germany, 2010.
141. Rescher, N. *Process Philosophy: A Survey of Basic Issues*; University of Pittsburgh Press: Pittsburgh, PA, USA, 2001.
142. Van Eemeren, F.; Grootendorst, R. *Argumentation, Communication, and Fallacies: A Pragma-Dialectical Perspective*; Erlbaum: Hillsdale, MI, USA, 1992.
143. Gunji, Y.-P.; Shinohara, S.; Haruna, T.; Basios, V. Inverse Bayesian inference as a key of consciousness featuring a macroscopic quantum logical structure. *Biosystems* **2017**, *152*, 44–65. [CrossRef]
144. Ames, R.; Hall, D. *Chapter 64 in Dao De Jing. A Philosophical Translation*; Ballantine Books: New York, NY, USA, 2004; pp. 177–179.
145. Bishop, R.; Brenner, J. Potentiality, Actuality and Non-Separability in Quantum and Classical Physics: Heisenberg's *Res Potentiae* in the Macroscopic World. *arXiv* **2018**, arXiv:1801.01471.
146. Luhn, G. The Causal Compositional Concept of Information. Part I: From Decompositional Physics to Compositional Information. *Information* **2012**, *3*, 151–174. [CrossRef]
147. Dodig-Crnkovic, G. Nature as a network of morphological info computational processes for cognitive agents. *Eur. Phys. J. Spec. Top.* **2017**, *226*, 181–195. [CrossRef]

 *philosophies*

Article

Contemporary Natural Philosophy and Contemporary *Idola Mentis*

Marcin J. Schroeder

Global Learning Center, Tohoku University, Sendai 980-8576, Japan; schroeder.marcin.e4@tohoku.ac.jp

Received: 6 July 2020; Accepted: 26 August 2020; Published: 1 September 2020

Abstract: Contemporary Natural Philosophy is understood here as a project of the pursuit of the integrated description of reality distinguished by the precisely formulated criteria of objectivity, and by the assumption that the statements of this description can be assessed only as true or false according to clearly specified verification procedures established with the exclusive goal of the discrimination between these two logical values, but not with respect to any other norms or values established by the preferences of human collectives or by the individual choices. This distinction assumes only logical consistency, but not completeness. Completeness (i.e., the feasibility to assign true or false value to all possible statements) is desirable, but may be impossible. This paper is not intended as a comprehensive program for the development of the Contemporary Natural Philosophy but rather as a preparation for such program advocating some necessary revisions and extensions of the methodology currently considered as the scientific method. This is the actual focus of the paper and the reason for the reference to Baconian *idola mentis*. Francis Bacon wrote in *Novum Organum* about the fallacies obstructing progress of science. The present paper is an attempt to remove obstacles for the Contemporary Natural Philosophy project to which we have assigned the names of the Idols of the Number, the Idols of the Common Sense, and the Idols of the Elephant.

Keywords: contemporary natural philosophy; *idola mentis*; scientific methodology; quantitative and qualitative methods; structural analysis; abstraction; complexity

---

Dedicated to Gordana Dodig-Crnkovic
who proposed the idea of the Contemporary Natural Philosophy Project

1. Introduction

Natural Philosophy or Philosophy of Nature has a long intellectual tradition with diverse ways of its identification as a style of inquiry and with the diverse interpretations of its role in the life of human collectives and in the individual reflection on reality. The presence of the qualification of Philosophy by the terms "Natural" or "Nature" does not make the concept easier to comprehend considering the long tradition of disputes about their meaning going back at least to Aristotle. Moreover, Natural Philosophy is the subject of this paper not so much because of its past, but because of its potential for the future of inquiry. This is the reason for the temporal qualification in the name "Contemporary Natural Philosophy" used in this paper and in the series of papers for which the present paper is intended. There is an increased interest in the revival and reconceptualization of Natural Philosophy as the means to adapt intellectual inquiries of reality to the challenges of complexity and of its consequences faced by science and philosophy [1,2]. Natural Philosophy emerged from the attempts to acquire universal knowledge of reality devoid of earlier divisions into separate realms of the Heaven and Earth consisting of separate essences long before the separation of the forms of this style of inquiry into the emancipated disciplines of knowledge. In this sense it can be considered a parent of the disciplines called Natural Sciences. Contemporary Natural Philosophy can be viewed as an attempt to reintegrate the vision of reality fragmented by the overload of complexity into a domain overarching Natural Sciences, but going far beyond these disciplines, and challenging conventional disciplinary divisions.

This paper does not have any ambitions to analyze the entire variety of past and present conceptualizations of the Natural Philosophy, its revival in the form of the Contemporary Natural Philosophy or to advocate for any specific choice for its identifying principles. Instead, its objectives are to look for that which is common in the diverse studies, directly or indirectly associating themselves with the naturalized inquiry of reality and to identify the fallacies which have to be eliminated or avoided, if we want to make this type of inquiry effective. In fact, the latter objective is primary and the former just sets the stage for the study.

The present paper is motivated by the view that the Contemporary Natural Philosophy can and should play the leading role in the process of developing an integrated vision of objective reality built with the use of a self-regulated by the feedback control methodology. This process does not have to be limited to the integration of the existing forms of scientific inquiry or to the organization of their accumulated results.

Although the choice of the name "Contemporary Natural Philosophy" for this gradually emerging domain of inquiry is far from being of primary importance, it can be justified by the affinity with the loose but identifiable tradition associated with the name "Natural Philosophy" in the intellectual history of humanity and, on the other hand, by the need to avoid the confusion with the existing fragmented, lacking cohesion, and dominated by external values and norms field of human activities conventionally called science. Although, in this paper, this conventional term will be used frequently along with the expression "scientific method", this is not an expression of the view that they refer to some clearly defined and uniform concepts, but rather a matter of convenience.

At present, Contemporary Natural Philosophy has the status of a project discussed in the series of papers presenting a wide variety of views and positions [1]. For this reason, its vision presented here is idiosyncratic and possibly temporary. However, no matter what its future shape will be, there is no doubt that its formation and development will require some adaptations and revisions of the methodologies which it inherited from Natural Philosophy and sciences. The present paper is intended as a preparation for these methodological transformations.

This is the actual focus of the paper and the reason for the reference to Baconian *idola mentis*. Francis Bacon wrote in *Novum Organum* about the fallacies obstructing science in its *statu nascendi*. The present paper is an attempt to remove obstacles for the Contemporary Natural Philosophy, categorized here, rather conventionally, as the Idols of the Number, the Idols of the Common Sense, and the Idols of the Elephant.

The reference to Francis Bacon does not mean an intention of the revival of Baconian philosophy of inquiry. It will become clear that the intentions of this paper are, in some cases, just opposite to those of Bacon. Its reason is the function of Baconian idols as a denouncement of the patterns and habits of human thinking which have to be eliminated for the purpose of achieving the authentic knowledge of reality. The function of the idols presented and discussed here is the same, but their specific characteristics are different and sometimes opposite to those in *Novum Organum*.

The triadic categorization of the Contemporary *idola mentis* which should be avoided in the development of the Contemporary Natural Philosophy is not intended to be comprehensive, exhaustive, or exclusive. It is as idiosyncratic as the vision of the new domain. After all, we are talking about the future domain of inquiry which is being discussed and developed. Possibly other idols will be identified in the future and they all may be re-categorized.

The selection of the three categories and of the examples of idols within these categories is dictated by my own experience from my mathematical-scientific research, from my teaching, and from my work on philosophical reflection. The main criterion for the inclusion of instances and types of fallacious reasoning into consideration in this paper was their hidden omnipresence in the present scientific, philosophical and educational practice, and their detrimental impact on this practice. Their elimination is of great importance for the development of the methodology for the Contemporary Natural Philosophy.

The last statement can generate disbelief and criticism of my inflated expectations. How can I know that some topics will be of great importance for the Contemporary Natural Philosophy before it is born and matured? The answer will be given later, but at this moment, I can only give examples of the topics which are addressed in the description of the idols. The Idols of the Number address misconceptions regarding the distinction between quantitative and qualitative methods in scientific methodology. The Idols of the Common Sense address misconceptions regarding the relationship between the formal conceptualization of elements of reality and the way we perceive reality. The Idols of the Elephant address misconceptions regarding the relationship between structural divisions of reality.

The use of the term "misconception" puts some normative load in these descriptions. Does it mean that the subject of the paper is tracing errors in scientific methodology? My preference is to talk not about errors, but rather about fallacies. Errors are deviations from some standards of precision or correctness which are not always available or known, especially in the context of the domain of study which is still in the process of development. Fallacies are more general, as they may be of the type of formal fallacies where the deviation from some standards (i.e., they may be errors), but also of the type of informal fallacies, where the issue is not the deviation from some standards, but in not meeting declared expectations [3]. Certainly, the latter form of fallacies is relative to the expectations and therefore, it requires some context. All idols studied in this paper have the context of the Contemporary Natural Philosophy, although some are formal and can be classified as idols independently from any context. They have the common feature of being based on typically hidden divisions assumed to be obvious and absolute in their status or in their mutual relations. The expectation which serves as the evaluative (negative) criterion is the goal of an integrated view of reality. The idols studied here are obstacles in building this view.

2. Contemporary Natural Philosophy

As it was emphasized several times above, the role of the description of the Contemporary Natural Philosophy is to provide a context for the main part of the paper about overcoming obstacles in its development. The strategy in this description is to minimize the restrictions on its further evolution. The vision of the Contemporary Natural Philosophy presented here is not a summary of its discussions carried out in other contributions to the subject or the result of a consensus in these discussions. It is more a proposal of the framework in which the project can proceed.

2.1. Motivations for Contemporary Natural Philosophy

The problem of the fragmentation of human intellectual activity was studied in many different contexts and perspectives from the level of the modern civilization to internal divisions of scientific disciplines. This is not directly the subject of the present paper, but the claim of the need for reintegration of the view of reality may generate the question about its justification. Thus, the most prominent points of the discussions will be reported shortly.

The global scale disaster of the Second World War with its unprecedented atrocities led to the recognition of the consequences of the fragmentation of human intellect distributed among the highly specialized experts who lost the vision of reality as a whole in all its interrelated natural and humanistic aspects. This recognition became the subject of the common interest and multiple disputes all over the world after the publication of the 1959 book *The Two Cultures* by C.P. Snow [4]. Snow directly blamed the split into Two Cultures, that of the humanities and that of science and engineering for the degradation of intellectual elites willing to engage or at least tolerate war crimes committed by their own authorities.

Actually, the concern about the threats to the values of a free society caused by the spontaneous election of specialized individual curricula by students who had free choice of courses motivated the administration of Harvard University to form the Faculty Committee on General Education already in 1943. The famous so called "Redbook" on General Education, with the univocally approved by the entire committee report and recommendation of the policy, was published two years later [5].

The difference between the two publications, by Snow and bythe Harvard Committee, was not in the diagnostic of the problem, but rather in their objectives. Both identified intellectual fragmentation as the main issue and both blamed the shortcomings of education for this fragmentation. Harvard Committee focused not only on the diagnostic, but also on the means to achieve the reintegration through the reform of the secondary and postsecondary General Education, which already existed, but which was ineffective in achieving its goals. Both publications promoted distribution of the subjects of study to provide graduates with sufficiently wide knowledge extending far beyond the subject of a more focused concentration of study. Recommendation from the Harvard Committee became the pattern for the entire American higher education, and both books influenced education in universities all over the world with some high and low levels of support through the decades. Whatever high or low points of General Education we can identify, it was clear that the goals of reintegration of Two Cultures were not achieved and that the cultures accelerated in their drifting apart.

Another form of the recognition of the problem of fragmentation together with the reflections on the methods of reintegration can be exemplified by the two books published by Edward O. Wilson, the first in 1998, *Consilience: The Unity of Knowledge* and the second in 2011, *The Meaning of Human Existence* [6,7]. Equally influential was the book *Mind and Nature: A Necessary Unity* published by Gregory Bateson in 1979 [8]. These are not the only books on the subject of the fragmentation of the scientific vision of reality and of the need for its reintegration, but the prominence of the authors generated a very wide resonance in the general public and among scientists and philosophers.

Not all works on the unification of science or its disciplines, such as physics, directly refer to fragmentation, but the fact that their focus is on the integration of subjects, methods or positions indicates that this fragmentation is problematic. For instance, Frank Wilczek, in his 2015 article, "Physics in 100 Years" does not lament lack of cohesion of physics, but considers different forms of the unification of physics as a most important aspect of the development of this discipline [9].

The discussion of the internal fragmentation of physics, apparently the most cohesive discipline of science, continued for years. It is generated and driven by the wide range of problems from the long chain of failures in the reconciliation between the Quantum Theory and the General Relativity Theory, through the failure to account for 94% of the universe mass and energy, popularly called the dark energy-matter, to more sensational, but not less serious problem of the failure to get clear judgment on the controversial work of the French brothers, Igor and Grichka Bogdanov, who at the turn of this century, defended their doctoral dissertations and published their papers in spite of the prevailing opinion that the content was pure nonsense expressed skillfully in scientific jargon [10]. In this last case, those who approved the degrees and accepted the papers and those who denounced them as fraud could not find the common criteria of evaluation.

Finally, an example of a comprehensive and in depth philosophical analysis of the issues related to the subject of fragmentation, but in the much more extensive context, can be found in the collective work of seventeen authors: *Stepping Beyond the Newtonian Paradigm in Biology: Towards an Integrable Model of Life—Accelerating Discovery in the Biological Foundations of Science INBIOSA White Paper* [11].

There is a natural and legitimate question whether the solution of the problem can be found within the existing framework of science by simply reestablishing more naturalistic standards for all forms of inquiry.

The two main sources of problems in the naturalistic positions giving science its primary role in the inquiry of reality are an unavoidable specialization of domains, disciplines, theories and the use of common sense as a substitution for the methodology of their re-integration. The fragmentation of science (actually, of the entire human intellectual activity) is a natural consequence of specialization as a method to overcome the complexity of all subjects of study. No individual can achieve even basic knowledge of all disciplines of inquiry. Progress in science requires an engagement of the large, specialized collectives and the division of labor within them. However, without any structurally organized system integrating the outcomes of specialized inquiry across the superficial disciplinary borders all scientific activities and the progress of work leads from the increasing complexity of the subjects of the study to the increasing complexity of the results of the study. This may be sufficient for solving more practical, technological problems which do not require a broad perspective; but without the large-scale integration, we cannot claim success in the conquest of complexity. For this reason, all philosophical discussions of naturalism and its relation to different forms of the scientific realism that refer to science or scientific method, understood as well-defined and consistent concepts or at least as clearly comprehensible ideas, are highly problematic.

At this point, a disclaimer regarding the negative impact of the fragmentation of science becomes necessary. The coexistence of competing approaches, conceptualizations, and results within science is its fundamental and necessary characteristic. Science is a discourse among diverse conceptualizations and hypotheses. The integration or reintegration of science is understood here, not as a process of achieving stable homogeneity, but rather as a dialectic and dynamic development of the common stage for this constant scientific discourse in the form of an overarching methodology for building a unified but evolving vision of reality. This vision has to evolve, as otherwise, we cannot expect any progress.

Traditionally, there was a common belief that this unification of the vision of reality can be achieved by the methodological reductionism to physics, considered to be the root of the tree of knowledge or by the ontological reductionism of reality to the subject of physical theories. Today, the positivistic view of sociology initiated by Auguste Comte as "physics of society," giving this particular discipline of science its distinguished position, is just a historical curiosity and physicalism is largely abandoned. Although some disciplines continue suffering from so called "physics envy," there are calls for the change of the "paradigm" through giving the priority to biology, cognitive science or some other domain of scientific inquiry, but they are not less naive than the other forms of the domain-oriented reductionism. The vacuum left by physicalism was never filled by a commonly recognized and rigorously developed methodology of integration. Moreover, physicalism remains in the scientific and philosophical discourses in the covert patterns of thinking, manifested openly only in the common and apparently devoid of any specific intentional use of the terms such as "physical reality" instead of reality, "physical space" in reference to the spatial aspects of reality, or "matter" understood as a synonym of mass. Instead of seeking a foundation for the integrated view of reality in the choice of a distinguished already existing discipline, a different approach was proposed in the series of Special Issues of the journal *Philosophies* [2]. This alternative approach of a comprehensive domain of study was given the name of Contemporary Natural Philosophy.

There is another issue which should be considered in the search for broadening the perspective of inquiries. From time to time, there are short-lived attempts to engage in the scientific discourse the alternative cultural traditions of inquiry. Probably the most prominent example is an explosion of discussions on the foundations of physics stimulated by the 1975 bestselling book, *The Tao of Physics* by Fritjof Capra [12]. The ideas adapted from the philosophical tradition of the East were used to provide a justification for formalisms such as that of bootstrap. The issue is that these encounters with alternative cultural conceptualizations of reality were just momentary, if highly amusing fashions and they never led to actual integration of the alternative methodologies of inquiry into scientific or philosophical methodology.

This intercultural intercourse is another, perhaps most difficult role of integration, which goes beyond the internal divisions of sciences. Even before any form of integration of the diverse cultural traditions within philosophical inquiry of reality is achieved or advanced, it is possible to utilize their experience for the purpose of a better understanding of the present scientific methodology, when it is viewed from the external perspective.

2.2. Philosophical Framework of Contemporary Natural Philosophy

No matter how the Natural Philosophy was understood in the past, it was always related to science in multiple roles of a predecessor, precursor, guide, or a tool for hermeneutics of scientific disciplines and theories. This makes science a natural, although not necessarily exclusive context for the discourse on the Contemporary Natural Philosophy as an integrated study of reality. In the following, this postulated form of inquiry will be addressed as already existing, although its identity is a matter of the idiosyncratic projection of the desired characteristics on the already existing but diverse tradition of the Natural Philosophy.

The present paper is not intended as a clearly formulated program for the future Contemporary Natural Philosophy, but rather as a study of conditions necessary for its design and implementation allowing progress beyond the present status of scientific knowledge or its philosophical interpretation. However, the possible diverse ways of understanding the tradition of Natural Philosophy, and of its relation to the presented here program of a new domain of inquiry, make some more specific explanation necessary. Our objective is to eliminate fallacies which are relative to the goals of the Contemporary Natural Philosophy. It is important to avoid confusion regarding its role and goals, so we have to start from disambiguation.

Contemporary Natural Philosophy does not have a subservient role with respect to science, but rather, it is a design for its extension, revision and revival as a style of integrated inquiry capable of overcoming the limitations imposed by complexity of reality. This style of inquiry should avoid the generation of complexity of its results, preventing their unification into a consistent vision. The revision should not be limited to the saturation of science with philosophy or philosophy with science. It is true that philosophy has an important role in science and it is sad and symptomatic for the deficiency of its present state that this role has to be explained and defended, as, for instance, in the opinion paper, "Why Science Needs Philosophy," written by a group of researchers and philosophers about the instances of the influence of philosophy on recent important developments in life sciences [13]. The fact that it is necessary to convince anyone about the value of interaction between science and philosophy is alarming, but it does not explain much about the desired direction of transformations in the style of inquiry. This, of course, does not lower the value of the attempts to promote the intercourse between science and philosophy, such as that in the mentioned paper.

Contemporary Natural Philosophy is not an addition of the layer of a "second order science" postulated under the influence of the "second order cybernetics" [14–17]. At the first sight, someone can be convinced by the statements made by the proponents of the "second order science" and may believe that its objectives are very close to those of the Contemporary Natural Philosophy, such as an internal control of the methodology of inquiry. Certainly, this form of control requires a very thorough study of the hidden assumptions influencing the outcomes of inquiry and we can easily agree with its necessity. However, a closer look makes it clear that the concern in the "second order science" is not about the methodological aspects of science, but rather about psychological determinants influencing scientists. This is clearly stated by Michael Lissack,

> The traditional sciences have always had trouble with ambiguity. To overcome this barrier, 'science' has imposed 'enabling constraints'—hidden assumptions which are given the status of ceteris paribus. Such assumptions allow ambiguity to be bracketed away at the expense of transparency. These enabling constraints take the form of uncritically examined presuppositions, which we refer to throughout the article as 'uceps.' [...] Second order science reveals hidden issues, problems and assumptions which all too often escape the attention of the practicing scientist (but which can also get in the way of the acceptance of a scientific claim) [17].

The most important fallacy of this view is in the claim about the "hidden assumptions which are given the status of ceteris paribus." The method of abstraction, more frequently called the method of idealization expressed as *ceteris paribus* (everything else equal) is not hidden at all, at least from the time of Galileo. It is the central tenet of scientific methodology seeking the reduction of complexity. The idea of the internal mechanism of the methodological control in the Contemporary Natural Philosophy is much closer to Heinz von Foerster's original concept of the "second order cybernetics" formulated in a much broader perspective of general systems [14].

Certainly, the intellectual experience of philosophy, including philosophy of science as well as the scientific studies in the subject of psychology and sociology of scientists and their organizations, or more generally, of human beings and their organizations, may be very useful in the Contemporary Natural Philosophy and its methodology, but they can contribute very little to the effort of reintegration of science. The studies of scientific activities of individuals or collectives in psychological or sociological perspectives are of great value for improvement of scientific organizations and they can contribute to the progress of science or inquiries in general. However, the greatest achievements in science and philosophy were frequently products of the work of exceptional, highly talented individuals who sometimes worked alone, removed from the influence of the mainstream intellectual trends or who rebelled against tradition. These individuals rarely could be analyzed in terms of the study of average members of human collectives. We can learn from their stories about creating the best conditions for fostering intellectual inquiries made by exceptionally talented individuals, but not about the desired directions of these inquiries.

More generally, Natural Philosophy is definitely not an equivalent of Philosophy of Science. They have very different methods and different objectives. The former attempts to study reality in a systematic way, engaging in some extent, the experience and methods of science; the latter has as its subject, science itself as a domain of human activity and its products. The central position of science in the project of Contemporary Natural Philosophy does not mean that science and its methods are of exclusive interest. Thus, while the naturalized epistemology of Willard Van Orman Quine is quite close to the spirit of the project, his famous and intentionally provocative statement, "philosophy of science is philosophy enough" [18] is very far from the vision of the new domain presented here.

Some parallels with Quine's thought or with the views of other philosophers should not be misleading. For instance, the present paper subscribes to the normative rule, "philosophy can, and should, make use of any of the forms of reasoning appropriate to scientific research," which is in perfect agreement with Larry Laudan's normative naturalism [19]. However, in the exact opposition to Laudan and his preference for "cognitive values" (scope, generality, coherence, consilience, and explanatory power) and "social values" (related to social processes of communication, negotiation, and consensus formation) over "epistemological values" such as truth, this paper defends the fundamental role of the concept of truth. Similarly, Quine's attempt to understand science exclusively from within the resources of science itself is too narrow to be acceptable as a principle for the Contemporary Natural Philosophy.

In agreement with the strategy to be as little restrictive as possible, there are only few characteristics of the Contemporary Natural Philosophy assumed here. Thus far, I presented more a "wish list" than actual description of the Contemporary Natural Philosophy. What would be the best description?

Contemporary Natural Philosophy is understood here as a pursuit of the integrated description of reality which is distinguished by the precisely formulated criteria of objectivity and by the assumption that the statements of this description can be assessed only as true or false according to clearly specified verification procedures established with the exclusive goal of the distinction between these two logical values, but not with respect to any other norms or values established by the preferences of human collectives or by the individual choices. Since the exclusive true-false distinction plays here a fundamental role, it has to be stressed that this distinction assumes only logical consistency, but not completeness. This means that completeness (i.e., the feasibility to assign true or false value to all statements which can be formulated) is desirable, but may be impossible.

Of course, Contemporary Natural Philosophy is a human product and as such, can be a subject of normative judgments at the level of meta-study. Moreover, its criteria of objectivity and its verification procedures evolve together with the collective experience of those engaged in the inquires and of the state of the overall vision of reality produced within the Contemporary Natural Philosophy. Thus, we can expect that the judgments of objectivity or truth from one stage of the inquiry may be reversed at a later time.

Additionally, the inquiries may involve the concept of probability in two ways. In one way, the probability of the truth of a statement in an inquiry can be considered as an expression of the epistemological use of the probability in inductive reasoning. This does not contradict the principle of the exclusive true-false logical values of acceptable statements, because the probability applies here to the knowledge of the logical status of the statements, not to the logical status itself. The other form of an engagement of probability theory can be directly in the statements about reality. In this case, the statement about the probability of some event within reality is about some aspect of reality and this statement can be assessed as true or false. Thus, here too, we do not have any inconsistency with the exclusive true-false values of the statements of inquiries. Some forms of the description of reality can have probabilistic form. The key point is whether this description is true or not.

It is easy to recognize the affinity of the Contemporary Natural Philosophy as presented here with Ilkka Niiniluoto's critical scientific realism [20], or with Michael Dummett's view of realism [21]. Both authors emphasize the importance of bivalence of truth-falsity as a necessary condition for realism, and this is in full agreement with the description of the Contemporary Natural Philosophy above. However, this paper is intended as a study of methods for a very broad and diverse direction of inquiry, and the emphasis on more specific understanding of realism or reality may defeat its purpose. For instance, the issue of the independence of reality from its human exploration (perception, cognition, empirical observation) is highly non-trivial and far from being established in the context of modern science, in particular of quantum mechanics, in which the state of quantum systems is not an observable and is dependent on the act of measurement. The only claim of the realistic doctrine acceptable here would be that the existence of any actual entity should be independent from our conceptualization of existence or from our will, but even this may require some qualification.

Even weaker forms of openly declared realism which do not refer directly to independence of reality from the inquiry may be too restrictive. The closest to the objectives of the Contemporary Natural Philosophy would have been the tenets of the realist liberal naturalism as presented by Mario De Caro:

> The tenets of realist liberal naturalism are: (i) A liberalized ontological tenet, according to which some real and non-supernatural entities exist that are irreducible to the entities that are part of the coverage domain of a natural science-based ontology; (ii) A liberalized epistemological tenet, according to which some legitimate forms of understanding (say, a priori reasoning or introspection) are neither reducible to scientific understanding nor incompatible with it; (iii) A liberalized semantic tenet, according to which there are linguistic terms that refer to real non-supernatural entities that do not form part of the coverage domain of natural science and are not reducible to those entities which do; (iv) A liberalized metaphilosophical tenet, according to which there are issues in dealing with which philosophy is not continuous with science as to its content, method and purpose [22].

However, the references to the concepts of "non-supernatural entities" or "non-continuity of philosophy with science" make this description of realism questionable and not acceptable for the characterization of the Contemporary Natural Philosophy.

Therefore, the issue of how to understand reality can be studied in a much more suitable context of objectivity than that of the independence from human observer or human observation, where objectivity is understood as invariance or covariance with respect to transformations induced by the change of observer or a reference frame. Probably the most suitable for our purpose is the concept of realism as the doctrine that the existence is separate or independent from the conceptions of it, which avoids commitment to the independence of the observer and the observed. It is true that such independence is preferred or even expected, but only the feasibility of knowing objective reality is postulated. After all, if we establish some form of rigorously defined objective criteria for objectivity (understood as invariance and opposed to subjectivity understood as the trivial invariance reduced to identity) [23], the definition of reality as that which satisfies these criteria is a simple consequence. While in the development of science (in particular physics) there was not much interest in the description of what reality is or what is real; the question about the criteria of objectivity was at the center of attention of scientific methodology, at least through the last four centuries.

Quite obviously, Contemporary Natural Philosophy is in direct opposition to Postmodernism and its denial of objective reality. For some contributors to the project, "Postmodern attack on Structuralism," which was probably the most important and most advanced attempt to reconcile natural sciences with the humanities, was a very strong motivating factor in the search for the revival of Natural Philosophy. From this point of view, Postmodernism can be used in the explanation of the ideas of the Contemporary Natural Philosophy as its antithesis.

In his 1979 metanarrative, *The Postmodern Condition: A Report on Knowledge*, Jean-François Lyotard initiated a crusade against metanarratives with the frequently repeated by others sentence: "Simplifying to the extreme, I define postmodern as incredulity towards metanarratives" [24]. The metanarrative against metanarratives is not the only self-contradiction of Postmodernism, but such contradictions seem not to bother the adherents of the revolt. The main topic of the book was a critique of metanarrative (or grand narrative) of science. Lyotard later admitted that his knowledge of science at the time of writing the book was negligible [25]. However, the critique, together with the commonly misunderstood Wittgenstein's idea of language-games (unfortunately, frequently interpreted without any basis in *Philosophical Investigations* that the use of the word "games" indicates that Wittgenstein dismissed any serious consideration for meaning) and with the openly expressed distaste for abstraction, led to the cult of the particular as opposed to general (power of individual event).

This confluence of ideas was promptly used against the ideas of Structuralism. This revolt against structuralism was deeper than just the anger generated by the perceived incomprehensibility and un-intuitiveness of scientific theories, and the limitation of the freedom of philosophizing by the requirements of the intellectual discipline imported from mathematics and science. Even stronger negative reactions were generated by the claims of the dismissal of apparently naturally existing chaos and disorder of the universe. Here, the opposition to the central ideas of Contemporary Natural Philosophy is the most direct and overt.

The irony of the intellectual history manifests here, once again. Lyotard's confession of his ignorance regarding science was sincere and his anti-scientific sentiment was very clear, but what he intended as a critique of science (e.g., of the lost commitment to the truth and the submissive conduct with respect to power and corporate interests) was actually an accurate critique of the social conditions in which scientific inquiry had to be conducted. Thus, the revolt was actually directed not against science, but against its corruption and ultimately, was in the name of science. There are some other points where the Postmodernist critique of science and scientific method could resonate among those who believe in the need for the transformation of science, however, the central tenet of the rejection of metanarratives in Postmodernism is irreconcilable with the fundamental commitment of the Contemporary Natural Philosophy to the search for the integrated understanding of reality.

3. From Baconian *Idola Mentis* to Contemporary *Idola Mentis*

The present revival of the interest in the Natural Philosophy reminds us of the situation in the past when Natural Philosophy started to emancipate from other forms of the philosophical inquiry and reflection at the beginning of the 17th century. This was the beginning of science before its fragmentation into specialized disciplines. At that time, it was necessary to reflect on the limitations of the Mediaeval Scholastic philosophical tradition and its sources in the philosophy of Mediterranean Antiquity. Baconian criticism of *idola mentis* was essentially nothing else but the "second order science" in the early 17th century format. Bacon was an equally adamant enemy of the involvement of abstraction or theory in the process of accumulation of knowledge, giving it only a secondary role of organizing the results in a more systematic way. This was an exactly opposite position to that of Galileo, who wanted to read the book of nature written in the language of geometry and who avoided the use of inadequate instruments when logical or mathematical reasoning was sufficient. For instance, he preferred to conceive the thought experiment of two stones tied together with a string as a justification for equal speed of falling objects over the falsely ascribed to him observations of falling stones thrown from the Leaning Tower of Pisa.

Baconian method was still firmly rooted in the passive observation in which he was very similar to his nemesis Aristotle, who strongly believed that this is the ultimate source of knowledge. Aristotle did not restrict himself to his own observations but accepted, in some cases, the accumulated knowledge from the observations of our predecessors reflected and preserved in the language. For Bacon, this would have been unacceptable. In Baconian vision of inquiry, the engagement of an observer's action was only in the "artificial" arrangement of the observed phenomenon outside of the usual context, but the observation itself was understood by him as a direct pathway from the perceptions of senses to the mind without any mediation of a theory or abstraction.

Galileo was aware of the importance of instruments, their construction, and of the influence of their inadequate precision. For this reason, he frequently replaced direct observation (of, for instance, free fall of objects) with the experiments involving manipulation of the observed system by an experimenter accompanied with the theoretical analysis of the settings and outcomes (as in the experiments with the motion of minimal friction objects on the inclined plane). Probably these important methodological differences were the main reason why Galileo made such important contributions to physics, while those of Bacon were almost exclusively to the organization of science and to the promotion of the idea of empirical methods.

While the direct contributions of Bacon to science, in general, and to physics, in particular, were of negligible importance, and his insistence on direct observation purified of any form of abstraction or theory was misguided, his reflection on the conditions for the effective ways of inquiry are still valuable now as they were in his time—of course, when we translate them into the language of modern science and consider them in the modern context.

The original four *idola mentis* denounced by Bacon were introduced in the following Aphorisms of *Novum Organum* [26]:

"XXXIX. Four species of idols beset the human mind, to which (for distinction's sake) we have assigned names, calling the first Idols of the Tribe, the second Idols of the Den, the third Idols of the Market, the fourth Idols of the Theatre."

"XLI. The idols of the tribe are inherent in human nature and the very tribe or race of man; for man's sense is falsely asserted to be the standard of things; on the contrary, all the perceptions both of the senses and the mind bear reference to man and not to the universe, and the human mind resembles those uneven mirrors which impart their own properties to different objects, from which rays are emitted and distort and disfigure them."

"XLII. The idols of the den are those of each individual; for everybody (in addition to the errors common to the race of man) has his own individual den or cavern, which intercepts and corrupts the light of nature, either from his own peculiar and singular disposition, or from his education and intercourse with others, or from his reading, and the authority acquired by those whom he reverences and admires, or from the different impressions produced on the mind, as it happens to be preoccupied and predisposed, or equable and tranquil, and the like; so that the spirit of man (according to its several dispositions), is variable, confused, and as it were actuated by chance; and Heraclitus said well that men search for knowledge in lesser worlds, and not in the greater or common world."

"XLIII. There are also idols formed by the reciprocal intercourse and society of man with man, which we call idols of the market, from the commerce and association of men with each other; for men converse by means of language, but words are formed at the will of the generality, and there arises from a bad and unapt formation of words a wonderful obstruction to the mind. Nor can the definitions and explanations with which learned men are wont to guard and protect themselves in some instances afford a complete remedy—words still manifestly force the understanding, throw everything into confusion, and lead mankind into vain and innumerable controversies and fallacies."

"XLIV. Lastly, there are idols which have crept into men's minds from the various dogmas of peculiar systems of philosophy, and also from the perverted rules of demonstration, and these we denominate idols of the theatre: for we regard all the systems of philosophy hitherto received or imagined, as so many plays brought out and performed, creating fictitious and theatrical worlds. Nor do we speak only of the present systems, or of the philosophy and sects of the ancients, since numerous other plays of a similar nature can be still composed and made to agree with each other, the causes of the most opposite errors being generally the same. Nor, again, do we allude merely to general systems, but also to many elements and axioms of sciences which have become inveterate by tradition, implicit credence, and neglect. We must, however, discuss each species of idols more fully and distinctly in order to guard the human understanding against them."

We can find in Baconian idols the reflections of the earlier philosophical thought. The Idols of the Den are not very far removed from Plato's Allegory of the Cave in his *Republic* which might have been the reason for their name. The similarity of Baconian Idols to the three centuries' earlier Roger Bacon's *offendicula* in *The Four General Causes of Human Ignorance (Causae Erroris)* forming Part I of his *Opus Majus* is very unlikely to be accidental [27]. Roger Bacon considered *offendicula* as the obstacles to acquiring real wisdom and truth, classified into four categories: (1) following a weak or unreliable authority, (2) custom, (3) the ignorance of others, and (4) concealing one's own ignorance by pretended knowledge.

In turn, we can easily recognize in Baconian idols the precedents of some major philosophical and scientific themes. For instance, the Idols of the Tribe refer to the bias common to all human inquirers, and coming out of the features of human senses and the ways they present objects to the mind. The issue whether we can overcome this bias of mediation and in what degree we can know reality was prominent in philosophical contributions of John Locke, David Hume, and most famously, Immanuel Kant and his followers. Another evidence for the continuing interest in the matters considered by Bacon is in the fact that the Idols of the Market can be almost directly translated into the Sapir-Whorf Hypothesis of the culturally determined features of the language influencing human cognition and therefore, shaping the way we comprehend reality [28].

The threat of being deceived by Baconian idols and the directive to adhere to the straightforward use of induction as the only tool of inquiry had its reflection in scientific contributions very different from the vision of Bacon. Even Isaac Newton, whose most important work *Principia* was very close to the style of, and openly patterned on Euclidean *Elements* and therefore, saturated with the purely theoretical style of inquiry in its axiomatic form, capitulated in the face of the mystery of gravitational action on the distance with his famous declaration of *hypotheses non fingo*:

I have not as yet been able to discover the reason for these properties of gravity from phenomena, and I do not feign hypotheses. For whatever is not deduced from the phenomena must be called a hypothesis; and hypotheses, whether metaphysical or physical, or based on occult qualities, or mechanical, have no place in experimental philosophy. In this philosophy particular propositions are inferred from the phenomena, and afterwards rendered general by induction [29].

Most likely, this *hypotheses non fingo* was an expression of his desperation. It was the only point where the laws of motion proposed by René Descartes, in his posthumously published in 1664 *Le Monde*, were superior in the strict adherence to the interaction on contact, easily defendable by the straightforward induction [30]. In the confrontation with the competing approach to the laws of motion, Newton probably did not want to open his work to the criticism of using clearly empirically non-testable explanations and he could not find the testable ones.

While the description of the idols in the *Organum Novum* is too simplistic to be used for the present scientific practice, they can help us to identify modern idols understood as habits of thought which can generate obstacles in inquiries of the Current Natural Philosophy, in particular, in its integrative role. There are three categories distinguished here: the Idols of the Number, the Idols of the Common

Sense, and the Idols of the Elephant, if we want to follow Bacon's style of giving names to categories of transgressions. The three categories are not entirely independent among themselves and not entirely independent from Baconian idols. The division into the triad of categories is purely conventional and rather a matter of convenience than a reflection of some deeper universal rules of human fallibility in the search for truth. They refer to the three different distinctions which are commonly but erroneously assumed to be obvious and absolute.

4. The Idols of the Number

This category is related to the common forms of misunderstanding of the role of mathematics in scientific or philosophical inquiries. The most prevalent but almost never questioned misunderstanding is in the belief in the fundamental distinction between quantitative and qualitative forms of inquiry. The latter, typically considered inferior, primitive or less "precise" is associated with qualification, i.e., with the partitioning of the set of objects according to their possession of some properties (or qualities). The former, apparently superior and more precise, is associated with quantification, i.e., with the assignment of a magnitude expressed as a number qualified by an occasional restriction that it has to represent a measure or count. This seems to be an ultimate dream of a scientist, and nobody asks the question when actually we know, when we know it. The focus is on whether other, preferably "independent" researchers confirmed the values, not on what numbers actually tell us.

It takes some knowledge of mathematics to realize the close resemblance of many claims of the "scientific" achievement in establishing the values of some magnitudes to the answer "42" given by the computer Deep Thought to the Ultimate Question of Life, The Universe, and Everything in the cult novel *The Hitchhiker's Guide to the Galaxy* by Douglas Adams [31]. There are many levels of misunderstanding in the fascination with numbers as a core of science starting from the most elementary, where the sources of misconceptions are simple errors, and the lack of mathematical education to quite advanced produced by another manifestation of fragmentation, when even very famous contributors to one sub-discipline of mathematics make statements well known in another sub-discipline as elementary errors.

4.1. What Do We Know When We Know the Number?

Thus, starting from the most elementary level, we have to eliminate the confusion of numbers and numerals which are their conventional, symbolic representations. Neither the arithmetic of natural numbers, nor any other theory can tell us whether $2 + 2 = 4$, $2 + 2 = 5$, or $2 + 2 = 11$, unless we fix the convention of numerical representation. Thus, Max Tegmark's statement:

> Modern mathematics is the formal study of structures that can be defined in a purely abstract way. Think of mathematical symbols as mere labels without intrinsic meaning. It doesn't matter whether you write 'two plus two equals four', '$2 + 2 = 4$' or 'dos mas dos igual a cuatro'. The notation used to denote the entities and the relations is irrelevant; the only properties of integers are those embodied by the relations between them. That is, we don't invent mathematical structures—we discover them, and invent only the notation for describing them. So here is the crux of my argument. If you believe in an external reality independent of humans, then you must also believe in what I call the mathematical universe hypothesis: that our physical reality is a mathematical structure. In other words, we all live in a gigantic mathematical object [...] [32],

in which he tries to summarize his central idea of The Mathematical Universe published elsewhere in a more elaborate format, is a surprising mixture of contradictory statements.

The first statement is in exact agreement with the view presented here in this paper, although with the usual unfortunate assumption that the term "structure" is already known, while it is probably the most frequently used but the least understood, and still lacking a sufficiently general definition concept in discourses on mathematics. It is followed by the statement involving "$2 + 2 = 4$" which suggests that this equality is true, but which is not formulated in the abstract language of mathematics (arithmetic), free from the dependence on convention and therefore, it is just a matter of conventional choice whether it is true or not. For instance, the first and the third equality above are true but in different conventions (decimal and ternary numerical systems, respectively). Then suddenly, we learn that, "That is, we don't invent mathematical structures—we discover them, and invent only the notation for describing them."

Yes, in the case when we write "2 + 2 = 4" we do not invent mathematical structures, but this is not writing a mathematical theorem. As Tegmark rightly stated at the beginning, "Modern mathematics is the formal study of structures that can be defined in a purely abstract way", and therefore, he admits that we define them, which for the purpose of simplicity of the language can be expressed that we create them or invent them. The statement "properties of integers are those embodied by the relations between them" is mysterious and difficult to analyze as it does not fit any standard use of the term "embodiment". Tegmark wrote earlier in the same article, "Here, I will push this idea to its extreme and argue that our universe is not just described by mathematics—it is mathematics" [32]. This pushing is definitely not very convincing but it demonstrates very well the dangers of the unfortunately frequent confusion of semantics with ontology. The fact that some statements have well-identified intentions, which is not exactly the case in the quoted passage where the meaning of some statements is unclear, does not entail their ontological status.

The confusion of numerals with numbers is only the first of many fallacies. To prevent it, statistical terminology of data requires that the main distinction between their quantitative type and qualitative type is that the former are expressed as numbers representing a count or measure, while the latter (possibly in the numeral form) represent the partition into disjoint classes (i.e., equivalence relations). The problem is that actually both types of data represent equivalence relations. In probability and statistics, it is clearly visible in the concept of a random variable when it is defined on a sampling space. The inverse images of the values of a random variable are simply classes of equivalence [33]. In the case of natural numbers (counts), this is more straightforward when we understand them as finite cardinal numbers defined by the equicardinality equivalence relation.

Thus, the engagement of numbers as values of counts or measures serves the purpose of the construction of equivalence relations on the set of objects of our inquiry, differing only in specifics from the qualitative analysis. This can be concluded from the casual reflection on physical magnitudes. They all have values expressed in conventional units, recently, mainly international SI units. The choice of the standard values is purely conventional and only the choice of fundamental magnitudes (called physical dimensions), although not free from some level of conventionality, is justified by physical theories. Here, it is easy to see that the particular value of the magnitude does not say anything about reality, but tells us about the equivalence class to which the outcome of observation belongs. Measuring the magnitudes is a tool to establish equivalence relations between the elements of reality. These equivalence relations in turn are involved in producing a wide range of mathematical structures such as partial order, topology, vector spaces, etc. At the same time, equivalence relations serve another purpose of lifting the level of abstraction when equivalence classes (subsets of the original set of objects) become the elements of the power set.

Finally, here is the key point of the methodology of the Contemporary Natural Philosophy understood as an integrative inquiry with the goal of reduction or elimination of complexity. Whatever is our understanding of natural or artificial intelligence, their most important feature is the ability to overcome the limitations imposed by the complexity of the environment. The primary tools for this purpose are the integration of information and abstraction [33–36].

The fallacies of the Idols of the Number do not discredit the importance of numbers in structuring our experience of reality. However, this importance has its source not in numbers themselves, but in structures which they form. At this point, it is necessary to refer to the next topic of the Idols of the Common Sense in which fallacies frequently arise from the illusion of obvious ideas.

Probably everyone (except mathematicians working in the number theory) believes that the concept of a number is obvious. School teachers sincerely believe that their introduction of the so called "real line" makes the concept clear and intuitive, when they draw on the blackboard a line, add to it an arrow on one side indicating the choice of one of the two possible choices of the linear order, mark two points indicating the location of 0 and the location of 1 and declare "To every real number different from 0 (we are done with it) corresponds exactly one point on the line which is on the right of point 0 if the number is positive, on the left if negative and which is in the distance from 0 equal to the absolute value of the number in consideration. Also to every point on the line corresponds exactly one real number identified as the distance from point 0 for points on the right side and its opposite for points on the left side of the line". Kids are happy that they can understand real numbers well and teachers are happy that they could give students a precise conceptual tool. After all, all concepts engaged in the construction of this structure are precise, clear and, at the same time, they are very intuitive by the reference to the association number-point. We do not need these nasty Dedekind cuts to understand numbers, is that not right?

The answer to the question is obvious: Wrong! The distance can be understood properly, only after we conceptualize real numbers and there is nothing a priori which we can use in the general case to determine where is the point with given distance to the point associated with 0; moreover, there is no way to determine what is right and what is left. This illusion of understanding is reflected in the popular belief that the numbers (real numbers) are very well understood and therefore, they provide the magic key to the proper understanding of reality.

The history of numbers reflects the entire intellectual history of humanity, but we cannot elaborate on it in this paper. Some moments of this history can explain the complications of a special importance in this study. Originally, in the European tradition with its sources in Greek philosophy, numbers were understood as those which we now call rational numbers (expressible, but not uniquely as proportions of integers), with the very clearly stated relational character, as they were derived from the arbitrarily selected geometric unit segment through geometric constructions. Numbers were proportions between geometric objects. For this reason, neither 0 nor 1 were considered numbers. It was the proof that the length of the diagonal in a unit square cannot be described by a number (i.e., rational number) that prompted search for the extension of the concept of numbers. The outline of the idea was provided by Eudoxus of Cnidos in times of Plato, but only in the second half of the 19th century did Richard Dedekind introduce a well-defined concept of real numbers based on what now we would call the completion of the linear order of rational numbers in terms of a Galois connection (i.e., mentioned above "nasty Dedekind cuts") [37]. By this time, the quantitative inquiry style was already established in the methodology of science, even if nobody really knew how to understand real numbers playing the central role in it. This should not be surprising, since even today probably less than 1% of people who use the quantitative methodology of science and who strongly believe in its superiority over the qualitative methods actually understand the concept of real numbers.

The importance of numbers as tools to describe the structure of order of the components of reality was already mentioned in the context of the extension of rational numbers to real numbers. However, it is only one type out of many structures which are generated by the association of numbers with the objects of reality. Moreover, in the text above, there was an emphasis on the recognition of the important distinction. The generation of the structures in the description of reality is not by numbers but by their structures. Theory of numbers is a theory of algebraic structures on sets of numbers. To appreciate the level of complexity and the importance of philosophical consequences of the structures of numbers, it is necessary to review some elementary mathematical facts.

4.2. Numbers and Their Structures

The presentation here will not require any previous knowledge of algebra beyond high school mathematics and the review is indispensable for the proper understanding of philosophical consequences of quantitative methodology. Apologies to mathematically educated readers who may decide to skip the presentation and proceed to its conclusion. If they read it, they may notice some restrictions of generality for the sake of simplicity (for instance, the consideration of algebras with n-arity of operations limited to at most 2) which prevents an unnecessary increase of complexity.

Algebraic structures or more formally general algebras are understood as sets equipped with one, two or many, sometimes infinitely many, operations. We can restrict our attention here to algebraic structures (general algebras) with two, one or zero arguments producing the result. The formal terminology is that these operations are binary, unary or nullary, respectively. The addition of numbers is an example of a binary operation: $a + b = c$ which takes two arguments a and b and gives the outcome c. Taking opposite number is a unary operation corresponding to addition $a \rightarrow a_+^{-1} = -a$ with the traditional notation a^{-1} for the inverse coming from the fact that the inverse for a non-0 real number a is its reciprocal $1/a = a^{-1}$. The nullary operation does not require any choice of arguments as its value is independent from arguments and consists in the selection of some constant element, for instance, the choice of 0 or choice of 1 which both have special roles of the neutral element as defined below. Notice that to define an operation on a set requires that for all arguments there is an outcome of the operation.

General algebras form an informal, traditionally and logically justified hierarchy (usually) starting from the concept of a semigroup $<S, \bullet>$ defined simply as a set S with a binary (i.e., two argument) operation $\bullet$ understood as a function from $S \times S$ to S which is associative, i.e., $\forall a,b,c \in S: (a \bullet b) \bullet c = a \bullet (b \bullet c)$. Whenever this is not confusing, we can drop the symbol $\bullet$ and write the juxtaposition ab instead of $a \bullet b$. The symbol $\bullet$ will be used only when the symbol of the operation without its arguments is necessary. Notice that the binary operation does not have to be commutative, i.e., in general, we do not require that $ab = ba$.

Thus, the only two conditions for a semigroup, below written in the simplified notation are:

➤ (no name as it is the universal condition for operation) $\forall a,b \in S \exists c \in S: ab = c$
➤ associativity can be written simply $\forall a,b \in S \exists c \in S: ab = c \; \forall a,b,c \in S: (ab)c = a(bc)$.

We define a neutral element e (the choice of letter is traditional) as an element satisfying the condition: $\forall a \in S: ae = ea = a$.

Semigroup with a neutral element e is called a *monoid*.

It is very easy to show that a semigroup can have, at most, one neutral element. Thus, we can say, "the neutral element e" when it exists. Both addition in real numbers (with the neutral element 0) and multiplication in real numbers (with the neutral element 1) define the structure of a monoid.

In a monoid with the neutral element e, we can define the concept of the inverse a^{-1} for an element a.

An element a^{-1} satisfying the condition: $aa^{-1} = a^{-1}a = e$ is called an inverse of a.

Once again, there is a very short and easy proof of a proposition: A monoid can have, at most, one inverse for each element. The respective inverses for addition in real numbers and multiplication in real numbers are for every a given by $-a$ and $1/a$, respectively.

This brings us to the most important general algebra in the entire mathematics defined as: A monoid in which every element has inverse is called a group.

The set of real numbers R with addition + forms a group $<R, +, 0, a \rightarrow a_+^{-1}>$. This group is called a commutative group, because for all real numbers: $a + b = b + a$.

There is a natural question: Is the set of real numbers R with multiplication a group? The answer is no, because 0 does not have a multiplicative inverse, since for every real number a $0a = 0$ and therefore $0a \neq 1$.

However, we can introduce the structure of a multiplicative group on the set $R^* = R \backslash \{0\}$.

The group $<R^*, \bullet, 1, a \rightarrow a_\bullet^{-1}>$ with respect to multiplication $\bullet$ on the subset $R^* = R \backslash \{0\}$ is commutative. Obviously, the subset R^* of R is closed with respect to multiplication, i.e., $\forall a,b \in R^*: ab \in R^*$ and $\forall a \in R^*: aa^{-1} = a^{-1}a = 1$ when $a^{-1} = 1/a$.

Now, we can consider an algebraic structure with two binary operations $+, \bullet$ on the set of real numbers R (for +) and R^* (for $\bullet$) $<R, +, 0, a \rightarrow a_+^{-1}, \bullet, 1, a \rightarrow a_\bullet^{-1}>$ where $<R, +, 0, a \rightarrow a_+^{-1}>$ is its additive commutative group and $<R^*, \bullet, 1, a \to a_{\bullet}^{-1}>$ is its commutative multiplicative group. This type of an algebraic structure is called a field if $\forall a,b,c \in S: a(b + c) = ab + ac$, i.e., multiplication is distributed over addition.

We can consider a general algebraic structure of a field $<K, +, 0, a \to a_+^{-1}, \bullet, 1, a \to a_{\bullet}^{-1}>$ (if no confusion is likely, we will write shorter: $<K, +, 0, \bullet, 1>$) defined not necessarily on real numbers but on a set K, where $<K, +, 0, a \to a_+^{-1}>$ is a commutative group (we say the additive group of the field) and K^* is a commutative group $<K^*, \bullet, 1, a \to a_{\bullet}^{-1}>$ where $K^* = K \backslash \{0\}$ (we say the multiplicative group of the field). We combine these two groups with the requirement that multiplication is distributed over addition: $\forall a,b,c \in K: a(b + c) = ab + bc$.

We will talk here only about a very few instances of fields and only about infinite fields (there are infinitely many of finite and infinite fields (!)). The most frequently used in applications are the field of rational numbers $<Q, +, 0, \bullet, 1>$, the field of real numbers $<R, +, 0, \bullet, 1>$ and the field of complex numbers $<C, +, 0, \bullet, 1>$. They form a sequence of the field extensions or (in reverse) of proper subfields:
$<Q, +, 0, \bullet, 1> \ll <R, +, 0, \bullet, 1> \ll <C, +, 0, \bullet, 1>$

The symbol $\ll$ indicates that what is on the left is a proper subfield (substructure, i.e., subset closed with respect to all operations of whatever structure is on the right).

The algebraic structure of a field $<K, +, 0, a \to a_+^{-1}, \bullet, 1, a \to a_{\bullet}^{-1}>$ (shortly written $<K, +, 0, \bullet, 1>$) can be found in many disciplines of mathematics and in many applications. The elements of a field K are what we call numbers or scalars, but this status is dependent not on individual elements but on the membership in the algebraic structure. It was already mentioned above that for the Ancient Greeks, numbers were elements of the field $<Q, +, 0, \bullet, 1>$ and it took more than two millennia to extend this field to the clearly defined field $<R, +, 0, \bullet, 1>$. For us, it is important that there are several important examples of infinite fields between the field of rational numbers $<Q, +, 0, \bullet, 1>$ and the field of real numbers $<R, +, 0, \bullet, 1>$, i.e., these fields form a chain of consecutive extensions or consecutive subfields of the field of rational numbers which in turn are subfields of real numbers.

Thus, the field of rational numbers Q is a proper subfield of the field of constructible numbers (numbers which can be constructed with the ruler and compass from the unit segment to the segment of the length equal to this number), which in turn is a subfield of the field A of real algebraic numbers (i.e., numbers which are roots of polynomials with rational coefficients), which is a subfield of the field of computable numbers, which in turn is a subfield of the field of definable numbers, which, finally, is a subfield of the field of real numbers R.

There was more technical reason for the further extension from the field of real numbers $<R, +, 0, \bullet, 1>$ to the field of complex numbers $<C, +, 0, \bullet, 1>$. This extension was dictated by the need to consider algebraically a complete field (i.e., a field in which arbitrary polynomial equations have solutions). The reason for the extension to complex numbers was more technical than conceptual, but it generates several philosophical questions. For instance, why do we accept only real numbers as values of physical magnitudes when, at the same time, we use most frequently the standard complex Hilbert space formalism in quantum mechanics? This is not a mathematical question which we can answer in a definite way as it is addressing intuitive preferences. However, the most likely answer is that the field of complex numbers loses the natural linear order of the field of real numbers. The field of complex numbers can be considered a two-dimensional vector space over the field of real numbers, and in two dimensions, we lose any meaningful linear ordering. The real number values of magnitudes introduce linear order in our description of reality. This feature is lost if we admit complex values.

All these fields $<K, +, 0, a \to a_+^{-1}, \bullet, 1, a \to a_{\bullet}^{-1}>$ (in short $<K, +, \bullet>$) in the chain considered above are defined on some proper subsets K of real numbers ($K \subseteq R$ and $K \neq R$) starting from the field of rational numbers Q. We can easily, and in the full agreement with our intuition, construct rational numbers forming the set Q from the integers in Z which in turn can be easily derived from the natural numbers in N. Of course, neither the set of natural numbers nor set of integers has the structure of a field with operations $+, \bullet$ as they lack multiplicative inverses. The field of rational numbers Q is the smallest field including all natural numbers.

Ancient Greeks thought that rational numbers are all numbers until they found that we need to look for an extension when we want to assign the length to the diagonal of a square with unit sides which today, we call the irrational number $\sqrt{2}$. The diagonal could be constructed with the use of ruler and compass, yet it was lacking the corresponding number. This deficiency of $<Q, +, \bullet>$ to represent geometrically constructible objects justified the need for an extension from the field $<Q, +, \bullet>$ to the field of what we call today, constructible numbers (more formally, the field of constructible numbers). Every number in this field can be associated with the length of the segment which can be constructed with the compass and ruler. However, there are also numbers like $^3\sqrt{2}$ which are roots of polynomials with rational coefficients (real algebraic field) which are not constructible. For instance, $^3\sqrt{2}$ is a solution of the equation $x^3 = 2$. Thus, when the elements of reality started to be considered in terms of equations, it was necessary to search for further extension. The next larger field is the field of computable real numbers which can be results of the work of a Turing Machine, i.e., the work of any computer. It is countable, so still much smaller than the uncountable field of the real numbers $<R, +, \bullet>$. The majority of real numbers are not computable. Even worse, the majority of real numbers are not definable. Between the field of computable numbers and the field of the real numbers, there is a countable field of the definable numbers. These are numbers which can be identified by a description in terms of logic and set theory. The uncountable majority of real numbers are not definable. There is no way we can identify non-definable numbers. They do not have any properties expressible in mathematical language which we could use to distinguish them.

The philosophical, e.g., ontological consequences of the recognition of these fields are enormous. It is difficult to accept the primary existence of the entity which does not have even, in principle, individual identity. Thus, how can we understand the identity of a real number which is not definable? Not only are these undefinable numbers in the majority of the set of real numbers, but the set of definable real numbers does not have a non-zero measure. Another question is: How to assess the school teachings about the real line which install in children the completely false sense of understanding of the real numbers and of the understanding of reality in terms of apparently superior quantitative methodology?

The Idols of the Number are not restricted to the fallacies related to numbers or to the fallacy of apparent distinction between quantitative and qualitative methodologies of inquiry. The name just refers to the most common fallacy involving numbers, but should apply to all forms of misuse, abuse and misunderstandings of mathematics as a tool of inquiry. The question of the ontological status of mathematical objects is not included in this category as it is of perfectly legitimate character. It is true that some contributions to this subject are biased by the Idols of the Number, but this should not prevent further studies of this subject.

4.3. Unreasonable Misunderstandings of Mathematics

There is one more type of fallacy, which at first sight, may look as clearly belonging to the Idols of the Number as they involve mathematics, but actually could be placed in the next category of the Idols of the Common Sense, in spite of the fact that they misguide not lay people, but highly respectable mathematicians or physicists. The example can be the naive reflection on *The Unreasonable Effectiveness of Mathematics in the Natural Sciences* by famous physicist Eugene Wigner [38]. Wigner mused on what he considered a mystery: "The first point is that the enormous usefulness of mathematics in the natural sciences is something bordering on the mysterious and that there is no rational explanation for it" [38]. He concluded his paper with: "The miracle of the appropriateness of the language of mathematics for the formulation of the laws of physics is a wonderful gift which we neither understand nor deserve. We should be grateful for it and hope that it will remain valid in future research and that it will extend, for better or for worse, to our pleasure, even though perhaps also to our bafflement, to wide branches of learning" [38].

The leitmotif of the paper is like a bewilderment of someone who, referring to his shooting skills, after watching the arrows in the center of the target's bull eye, forgot that the arrows were shot first and only after this, were the concentric circles drawn. Wigner's surprise is one more piece of evidence for the fragmentation of science which started to be considered as the normal state. In the past, there was no separation of mathematics and physics, therefore, the work on physical theories was not different from the work on mathematical problems. There is no surprise (although apparently for Wigner there is) that informal, intuitive associations between different domains of scientists' activities acted as cross-pollination between mathematics and physics, even if, very often, the formal association might have been never considered or achieved. Mathematical theories frequently went much beyond the interests of physical theories and the connection was lost.

Wigner's article could have been just an amusing anecdote about an absent-minded famous physics professor who suddenly realizes that instead of doing his job in physics, he is doing mathematics. However, the sensational title of the paper and the Matthew effect caused a lot of damage by creating a frequently invoked false mystery. It is hard to believe that *Unreasonable Effectiveness of Mathematics* was written by the founder of the studies of symmetry and group theory in quantum mechanics, who received the 1963 Nobel Prize in Physics for his contributions to the theory of the atomic nuclei and elementary particles through the discovery and application of fundamental symmetry principles.

An explanation of "[t]he miracle of the appropriateness of the language of mathematics for the formulation of the laws of physics [...]"can be found in the works of other giants such as Hermann Weyl [39] or the 1977 Nobel Prize in Physics laureate Philip W. Anderson [40], who gave the answers to the role of mathematics in physics and other disciplines exactly in terms of symmetry and group theory. The apparent miracle turns out to be just an expression of the identity of mathematical and physical theoretical inquiries as summarized in a sentence from Anderson's famous article "More is Different": "It is only slightly overstating the case to say that physics is the study of symmetry" [40]. Yet the title of Wigner's article became a meme which persists in confusing lay people. Whatever damage has been done, the case is an excellent example of the dangers of the Idols of the Number which include the belief in the reality of a strict demarcation line between mathematical and physical inquiries.

5. Idols of the Common Sense

It is again necessary to start this section from a disambiguation. Here, too, we have to be aware of the possibility of the confusion caused by equivocation. Common sense has two separate, although convoluted, traditions of study and associated with them are multiple ways of understanding this expression. Common sense in the understanding presented by Aristotle in *De Anima* is a capacity to identify shared aspects of things. Various expressions involved in the analysis of this capacity in humans and animals were later subsumed in the later translation into the *"common sensibles"* (and in modern psychological terminology called "binding"). Aristotle excluded existence of the sixth sense (although sometimes he addressed this capacity as the first sense), but rather considered the common sense a faculty by which *common sensibles* are perceived together as a single object.

Further evolution of this synthesizing faculty was long and too complex to be presented here, as this way of thinking definitely does not belong to the Idols of the Common Sense. Actually, it should be the subject of intensive studies within Natural Philosophy as a main tool for its integrative functions. An extensive study of common sense as the capacity to integrate information was published by the present author elsewhere [41].

There is another use of the expression "common sense" as a skill of using everyday experience common to all people from a more or less culturally homogeneous community in making decisions or normative judgments including judgments of the truth or falsity of statements. These type of skills are usually associated with "streetwise wisdom". Very often these skills are transmitted by language or learned by observation in the social environment. They may be of great practical value and they may be, in some situations, the only means to reduce complexity of the environment, i.e., they are necessary for everyday intelligent behavior. One of the main objectives of robotics and AI study is to develop in artificial systems the capacity of such common sense. Thus far, this objective has never been achieved.

Yet, we have to be careful about engaging both types of intuitive capacities in situations when the environment is very different from the environment in which intuitive skills have been acquired. Even more dangerous is mixing the intuitive and rational methods of inquiry involving higher levels of abstraction.

5.1. Beware of What Escapes Awareness

Idols of the Common Sense represent fallacies resulting from making conclusions based on individual, everyday experience, unaided by any systematic methods of critical thinking about the matters far removed from this experience. However, the origin of these type of fallacies is the result of the negligence of the recognition for both rational and irrational forms of inquiry and resulting confusions. When we ignore the role of the intuitive capacities as primitive and not deserving attention, they take over the functions of rational capacities and confuse them. In the presentation of the Idols of the Number, the central fallacious forms of inquiry were generated by the illusionary distinction between quantitative and qualitative methodologies of inquiry and the neglect of structural analysis accompanied by misunderstanding of mathematics and its role as a tool of inquiry. In the Idols of the Common Sense, the central confusion regarding the complex relationship between the rational and intuitive forms of inquiry, in mixing their analyzing and synthesizing roles is accompanied by the neglect of logic.

The distinction and relation between the rational and intuitive forms of inquiry was studied in my earlier publications [33,41,42]. For the purpose of this paper, it will be sufficient to consider the distinction between the inquiries involving the language-based reasoning organized and controlled by logic and the engagement of the human capacities to organize perceptions which escape linguistic and logical control. The most important capacity of the second type is our ability to integrate information into indivisible units which, in the rational form of inquiry, is called an "object". The examples of the interaction between the two forms can be found in the presence of the word "thing" in Aristotelian writings, which he never tried to explain or to define, or in the struggle to conceptualize the notion of a set (Cantor, Husserl and many others) which ultimately was abandoned by giving the notion of a set the status of a primitive concept.

The further consequences of the Idols of the Common Sense are especially detrimental for the study of the complementary objective and subjective forms of inquiry leading to the belief in their opposition and in the dominant and exclusive role for the former. The distinction here was explained very briefly in Section 2.2 in the terms of invariance, but was extensively discussed in my earlier publications [23].

If we want to study Contemporary Idola Mentis for the purpose of preventing errors and fallacies in the Contemporary Natural Philosophy, we have to avoid unjustifiably rigid rules and exclusive restrictions to the existing methodology of science. There is nothing wrong in the study and development of methodologies engaging human intuition and its capacities. There were many highly recognized mathematicians and physicists (e.g., Henri Poincare) who openly declared the primary role of their intuition in their achievements.

Yet, the collective experience of mathematicians and physicists provides examples of the abuse of what was considered a systematic use of intuitionistic methodology, in particular, the refusal to accept the excluded middle rule of logic. The most notorious was the abuse by Leopold Kronecker, of his power of being the editor, to veto the publication of works submitted by the founder of set theory, Georg Cantor, on the grounds that this was not mathematics. We are not concerned here about the development of clearly formulated programs to modify logic or other tools of inquiry, as long as they follow the rules of evaluation of intellectual activity and are not just expressions of individual belief or personal preferences. So, the excesses of intuitionism, even if being harmful, do not belong to the Idols of Common Sense. For this qualification, it is necessary that the common sense deviation from logical or systematic, methodological rules is without any awareness of it. The actual Idols of the Common

Sense are the cases when the intuitive forms of inquiry developed in the familiar environment from everyday experience encroach on the functions of rational capacities.

The classical example of the fallacy belonging to the Idols of the Common Sense has its own name, "Linda the Bank Teller". Amos Tversky and Daniel Kahneman, studying extensional and intuitive reasoning, created a story about a fictitious character called Linda for the participants in their research [43]: "Linda is 31 years old, single, outspoken, and very bright. She majored in philosophy. As a student, she was deeply concerned with issues of discrimination and social justice, and also participated in anti-nuclear demonstrations. Which is more probable? 1. Linda is a bank teller. 2. Linda is a bank teller and is active in the feminist movement." More than 80% of participants in the research chose answer 2.

We should not be surprised at the above. Probability theory and logic are notoriously counterintuitive. This is a natural consequence of the differences between competences of the rational and intuitive capacities. It is significant that the most confusing for untrained people are problems related to conditional probability (in particular Bayes Theorem) and to inferences involving implication. However, even the use of simple connectives such as "and", "or" turns out to be problematic for students, if they cannot use Venn diagrams (i.e., set theoretical representation).

5.2. Definition of the Definition

"Linda the Bank Teller" seems harmless, but actually, similar fallacies are surprisingly common in philosophical and scientific discourse, becoming a large obstacle in mutual understanding. This is a part of a larger problem in the context of much more serious misunderstanding of the concept of a definition and definability.

Once again, we can see the danger of equivocation, which can be identified as a main source of the Idols of the Common Sense. There are many different meanings of the word "definition" when it is qualified by some adjectives. Definitions always serve the identification of something. For instance, the dictionary definition serves the purpose of the identification of the standard use of words by finding their synonyms or synonymic expressions possibly with many words, typical paraphrases, or by providing contrast to similar words used in a very different way. Dictionary definitions are circular out of necessity, but also intentionally, as it is assumed that someone may know some words but not the others. Another example is an ostensive definition identifying objects by directly pointing at them.

There are multiple other "definitions" serving different objectives. However, in the context of philosophical or scientific inquiries, there is only one concept of a definition within formal logical methodology called genus-species definition. The tradition of this type of definition goes back to Socrates, but the works of Aristotle made it the central tool of philosophical methodology. Formal definitions of universals, i.e., terms with general meaning addressing multiple individuals, became necessary for the development of syllogistics as a methodology of reasoning. After Aristotle, the unqualified term "definition" always refers to genus-species definition. All other definitions, which are diverse forms of identification of a variety of objects or relations, require qualification.

Aristotelian concept of the *genus-species* definition referred to the partial order of universals according to their level of generality. The pair of universals was considered in the genus-species relation if every instance of the latter was an instance of the former. Thus, whenever we have that every A is B, A is species and B is genus. Of course, in the much later adopted biological taxonomy, the meaning of the terms "genus" and "species" changed as names of the specific consecutive levels of such order. In this partially ordered structure of universals, going in the directions of *species* was going in the direction of increasingly smaller classes of individuals, while going upward in the *genus* direction led to increasingly larger classes. Aristotle did not consider individuals being universals, but we could modernize the description of the structure of universals by considering individuals as atoms of the partially ordered set of universals.

To define a universal (*definient*) we have to identify its *genus* (one of them, but preferably *genus proximus*, i.e., the nearest of all genera). Then we have to provide *differentia*, i.e., we have to provide the difference between the universal which we want to define and all other species of this genus. The classical example of the definition was: A human (*definient*) is an animal (*genus*) which is rational. Being rational was the differentia which made the distinction between humans and other animals. The genus and differentia formed the *definiens*. Of course, we have to be able to identify a genus before we can use it in the definition. Thus, we had to have its definition ready, or we have to give it the status of a category, i.e., primary, undefinable, universal identifiable only with the use of intuition. Aristotle selected his own categories and in the millennia to come, philosophers formed their own selections. The border between the rational and intuitive forms of inquiry is exactly at this point. The selection of categories is beyond our rational capacities and it is left to our intuitive capacities.

The only difference in the modern formation of a conceptual system is that we do not feel obliged to start from the selection of all categories, but for a particular theory, we choose its primitive concepts (which we do not define) and we formulate a set of axioms as a priori true sentences characterizing the primitive concepts. From the axioms, we derive the truth of all theorems of the theory using valid logical inferences. Of course, the truth of theorems is conditioned by the assumed truth of axioms. To reduce the complexity of the statements, which we want to prove, and give them the status of a theorem of the theory, we can (and actually we do) define derivative concepts using the process of the definition starting with primitive concepts as *genera*. Later we can use, in *definiendum*, already defined concepts as *genera* for consecutive definitions.

It is important that, in principle, we do not have to define additional concepts, and by the cost of extreme complexity of the statements, we could develop an entire theory using only primitive concepts. Here, we can recall another contribution of Eugene Wigner, this time to the Idols of the Common Sense, when, in his paper, "The Unreasonable Effectiveness of Mathematics in the Natural Sciences" mentioned in the context of the Idols of the Number, he wrote, "Mathematics would soon run out of interesting theorems if these had to be formulated in terms of the concepts which already appear in the axioms" [38]. Of course, in this case there is not much damage as a disproval of this false statement can be found in any introductory textbook to logic. This just shows the dangers of misconceptions which can make even laureates of the Nobel Prize victims of the Idol of Common Sense. This also provides the justification for the inclusion above regarding the elementary explanation of the concept of definition.

This does not mean that definitions do not have practical importance. Not only do they direct the attention of the philosophical or scientific community to a particular direction of research but also, they simplify both reasoning of the author and its reception at the other end of communication. There is a good analogy in the use of the higher level programming languages. Of course, in principle, every program can be written in the machine language, but in such form, it would be practically incomprehensible to other human programmers. Higher level programs use defined subroutines which have a short name easily comprehensible to human programmers, and when they use these names in programming, the reversal of the names to machine language is performed by the compiler.

Definitions are not true or false. They are conventional tools reducing complexity of the language, but they are still conventional. In arithmetic, we write 5, not S(S(S(S(S(0)), but without the convention of writing digits in some particular way, we cannot understand the meaning of 5 using only arithmetical theory, which describes the primitive concept S(n) in terms of a recursive scheme.

The typical problems arise when the process of defining concepts, which is a syntactic procedure, is confused with semantics. The fact that we provide a definition of a concept does not tell us anything about the relevance of this concept, even if it is formulated in a perfectly correct way. We did not create anything new. We just eliminate a concept by reducing it to other concepts. This is actually an expression of the two main conditions for proper definition called "eliminability" and "noncreativity" [44]. Herbert Simon writes about them: "These criteria stem from the notion, often repeated in works on logic, that definitions are ('ought to be'?) mere notational abbreviations, allowing a theory to be stated in more compact form without changing its content in any way" [45].

There is extensive classic literature on the modern logical theory of definitions and definability with the particularly highly respected and renowned contributions of Alfred Tarski and Patrick Suppes [44,46]. The form of a logically correct definition is very well established and does not require much more study. The actual subject of the theory of definitions and definability is the transition between deferent theories developed in not necessarily the same conceptual framework of primitive concepts and axioms. This subject is beyond the scope of the present paper. After all, the most important lesson from logic about the concept of a definition is that Humpty Dumpty was right in his teaching Alice about the meaning of words: "When I use a word, 'Humpty Dumpty said, in rather a scornful tone,' it means just what I choose it to mean—neither more nor less" [47].

The idol which Linda the Bank Teller manifests is a quite frequent form of unintended and undesirable restriction of the scope of the concept by adding either additional differences or by adding additional axioms for the axiomatic theories based on the primitive concepts. Very often, authors who are not satisfied with the too narrow scope of the existing definition add to it additional conditions or comments, not realizing that this will never make the concept more general, but usually the effect is exactly opposite.

The logical definitions may not be sufficient for the purpose of theories describing a part of reality in terms of active engagement of observers. In this case, very often, operational definitions are used. They describe, for instance, how to construct the object of study through practical manipulations of the environment. This, of course, is very different from the presented before theoretical definitions. However, the difference can be eliminated if we include a theory of these operations into the more comprehensive theory of the studied fragment of reality. Once we have a theory of operations (for instance, empirical procedures) the operational definition can be formulated in the purely logical form.

Francis Bacon wanted to eliminate the intervention of theoretical reasoning in the form of theoretical description of experimental system, but in the perspective of modern science, his dream is impossible. Even if we could avoid the use of any experimental equipment (we know that we cannot) and restrict all inquiries to direct human observation based on sensory experience, our body is an experimental system and the functioning of our senses cannot be ignored.

The importance of the process of the formulation of the definitions for the concepts forming the conceptual reference frame can be seen in the eternal disputes on subjects, formulated as a question "What is...?" For instance, the concept of culture has been discussed since the 19th century by anthropologists, linguists, scholars of intercultural communication, etc., without ever reaching consensus. Already in 1952, Alfred L. Kroeber and Clyde K.M. Kluckhohn summarized, in a critical review, 164 earlier definitions, adding their own [48]. Arthur Lovejoy, in 1927, studied 66 ways in which the word "nature" has been understood in the context of aesthetics [49,50]. Raymond Williams based on the variety of definitions for nature, calledit "perhaps the most complex word in the language" but he was not aware that no philosophically non-trivial concept has commonly accepted unique meaning [51,52]. Even the concept of meaning has diverse meanings. The classical book of Charles Kay Ogden and Ivor Armstrong Richards, *The Meaning of Meaning* published in 1923, distinguished 16 different ways in which meaning is understood [53].

There is nothing wrong with the diversity of definitions. Actually, this diversity is just an evidence for the relevance. The problem is that the vast majority of so called "definitions" are not definitions at all from the point of view of logic. Quite a typical fallacy is that establishing of a quantitative magnitude is sometimes considered a definition of a concept.

The classical example of this fallacy of taking the definition of a mathematical formula for some magnitude as a definition of the concept is "Shannon's definition of information" which supposedly was written by Claude Shannon in his famous 1948 paper, "A Mathematical Theory of Communication" later published together with Warren Weaver in book format [54]. Shannon claimed to be interested in the fundamental communication problem of reproducing, at one point, either exactly or approximately a message selected at another point. In his paper, he formulated a mathematical concept of entropy characterizing probability distributions and wrote in Section 6, with the title, Choice, Uncertainty and Entropy: "Quantities of the form H = $-\sum p_i \log p_i$ (the constant K (omitted in the formula, m.j.s) merely amounts to a choice of a unit of measure) play a central role in information theory as measures of information, choice and uncertainty" [54] (p. 20). There is not much more directly about information in this historical paper, yet it is considered that Shannon defined here "information". It is clear that the two idols, of the Number and of the Common Sense, are responsible for this opinion. The former prompts people to believe that something expressed as a number giving value to some magnitude must be an entity. The latter idol just obscures the meaning of the definition as a concept.

Even when all definitions of some diverse attempts to define a concept are correct, the Idol of the Common Sense may prevent their effective use. The disputes on the definitions are often performed as if it was a matter of truth or falsity or of correctness. The definitions of concepts (if correctly formulated) can be evaluated exclusively on the adequacy of the theory which they serve, not by the form or content of the definition. If the theory (i.e., its syntactically true sentences or claims) describes objects of reality in the way which can be empirically confirmed, then we can consider the definition useful, but, of course, not true.

Another possible criterion of the evaluation of a definition can be formulated through the analysis of its conceptual framework (concepts involved in the *definiendum*). If a definition gives a wide range of relations with other relevant concepts, then this gives the evidence of its potential value, but this, too, can be assessed only by the analysis of the theory and its consequences. We have to remember that a definition of the concept is basically a selection of already defined or primitive concepts, something which metaphorically we could describe as a "conceptual system of coordinates". The same way as coordinate systems may be convenient or not is less important than finding the rules which govern phenomena framed by the coordinate system.

6. The Idols of the Elephant

The Idols of the Elephant can be easily recognized because of a well-known parable of "The Blind Men and an Elephant" accompanying many threads of Indian philosophical tradition, and going back before its first historical appearance in the Buddhist texts more than two millennia ago. The parable is now well known all over the world. A group of blind men tries to learn what a large object is, in their way. They use their tactile sense, but without having ability to see, they cannot compare and synthesize their individual experiences derived from touching small portions of the object. This may look like a too simplistic metaphor of the fragmented vision of reality provided by science. Certainly, the parable is of high relevance for Natural Philosophy as an integrated system of knowledge of entire reality, as it suggests that we should look for some form of sense of sight (or insight) to achieve integration.

Actually, not all the Idols of the Elephant are as obvious as that represented by the ancient parable. The other idols which prevent us in achieving our goal of integrated vision may not be like that in the parable, where the men are aware of their handicap. As in the cases of other idols, we may not be aware of our handicaps.

The division into the types of idols is not exclusive and not straightforward. In the description of the Idols of the Common Sense, the central fallacy was related to the relationship between the rational and intuitive forms of inquiry which have consequences for the relationship between objective and subjective forms of inquiry. However, this distinction can be associated with the Idols of the Elephant, too. Objectivity can be viewed as intersubjectivity, i.e., invariance with respect to the transition between the different human knowing subjects rather than the different, more general and not necessarily human observers or reference frames.

Let us change the parable and let all blind men touch at the same time, the same part of the elephant. Why should we expect that their reports should be the same? Each of the men has a different experience in touching the object of their environment. They may have different levels of sensitivity in tactile perception. Finally, they may have different skills in expressing their perceptions. Thus, we have to avoid oversimplified one-dimensional conclusions about the value of the collective inquiry.

The danger here is in the habits perpetuated by the language, promoting the view of the complementary objective and subjective forms of experience and inquiry falsely considered as exclusive, contradictory, competing, and requiring the dominance of one form over the other. The priority is usually given to the former, objective form. This can be seen in the normative character of the terms, "objective" (good), "intersubjective" (neutral) and "subjective" (bad) in everyday language.

Usual studies of objectivity are focused on preventing bias coming from the conflict of interest present in social life or from psychological determinants such as the trait ascription bias exhibited in a tendency to describe own behavior as flexible, adapting to the situational factors while the behavior of others is by ascribing fixed dispositions to their personality. Objectivity of science is expected to be achieved by the requirement of the judgment of many disinterested and independent reviewers. Sometimes, objectivity is considered in more abstract terms of independence of the evidence from that or whom it serves. Peter Kosso considers more general description"Objective evidence is evidence that is verified independently of what it is evidence for" [55].

We could see in the discussion of the Common Sense that misunderstanding of the concept of definition may generate difficulties in coordination of individual inquiries and formulation of a consistent vision of reality. However, this looks like a matter of communication between the blind men in the parable—but is the deficiency of communication the main problem? The problem is rather in the lost sight, i.e., missing tool of integration at the level of the acquisition of knowledge.

Certainly, it is very important to establish social mechanisms eliminating the influence of external factors and interests on the inquiries and their outcomes. Equally important is to foster good communication coordinating and integrating collective forms of inquiry. However, the most dangerous Idols of the Elephant are highly non-trivial and difficult to identify and control. They may not be related to the problems of coordination of inquiries performed by different individuals. They may put the obstacles on the path of inquiry of an individual inquirer.

In this paper, only one example of such non-trivial form of the Idols of the Elephant will be considered. This is a tendency to avoid the recognition of the hierarchic character of reality or in the attempts to giving, without any justification, the privileged status of reality to one particular level of this hierarchy. The blind people in the parable experience separate parts of reality (the elephant) in this parts' geometric separation. Each of the blind men is touching different parts of the surface of the elephant. It is still quite easy to reassemble the picture of the animal by gluing together fragmented images. We can consider yet another version of the parable of the (rather science fiction) ability to penetrate the body of the elephant to different depths. Once again, their reports will be different.

Reality can be analyzed from another perspective of having multiple levels of the collections of its components. In the mathematical language of the set theory, these levels can be constructed with the concept of a power set. We start from some set S which forms the first level of the hierarchy. Then we consider the set of all its subsets, which is called the power set of set S. Of course, we can form the power set of power set and the constructions can continue forever. On the other hand, all (usual) set theories do not have any separation of sets and elements. All objects of these theories are sets (no matter how strange it may seem to non-mathematicians). The concept of an element is relative. One set, let us say, set x, can be an element of another set y, which for the purpose of simplicity is expressed as "x is an element of y" ($x \in y$), but it does not give x and y different status. It is just an expression of the relationship between sets. This gives us the possibility of the infinite downward hierarchy.

We already know from physics that moving from one level of this hierarchy to another requires some change of the conceptual tools for the study of the collective phenomena which do not have any meaning at the lower level. We have examples of emergent phenomena whose description or prediction cannot be derived from the lower level. These are well-known ideas. Little less known or recognized is the fact that for the description of symmetry, the most important concept in many disciplines of science and humanities, we have to consider three levels of this hierarchy. If we want to consider higher order symmetries, we have to consider more levels [23,56,57]. Thus, there is no reason to think or to believe that these levels are just a creation of the human mind. At least, we should consider this hierarchic structure of reality a central subject of study and we should assess its ontological status based on the results of this study.

Now, the Idols of the Elephant appear here because there is a consistent tendency in many domains of inquiry to flatten the vision of reality. Traditionally, this tendency was expressed in reductionist forms of physicalism. In this position, we can only consider as real, one level of the hierarchy. Originally, this distinguished level was a stage of the set of points of space-time in which atoms or point masses were actors. Later atoms were replaced by elementary particles and the empty stage of points was equipped with the assigned to them vectors of the fields. All collectives of the higher level were just abstract creations of our inquiry without any right for independent existence.

Both the physicalist and reductionist position have lost attractiveness and are currently retreating, mainly under the influence of the reflection on the forms and mechanisms of life. However, the tendency of flattening reality remains, for instance, in the form of a variety of doctrines of structural realism initiated by John Worrall in 1989 [58]. The change in this direction consists in giving exclusive or, at least, primary existence to the second level instead of the first. Whichever level we choose, it may be the perspective of an individual blind man. To avoid this type of the Idols of the Elephant, we should wait for giving the priority to any particular level. If (for some unlikely reason) there is a reason to prioritize some levels over the other, this has to be justified by an empirically testable explanation and justification. In the absence of this justification, the hierarchic architecture should be retained until demonstrated otherwise.

The "flattening" tendency can be identified not only in natural sciences, physics, or philosophy. In some sense, it can be identified in mathematics, too. The shift in mathematics with some analogy to the shift in philosophy towards structural realism can be identified, for instance, in the category theory. In the admittedly oversimplified summary of the category theory, it can be described as an attempt to eliminate the first level. We give the priority to morphisms acting on objects. In the traditional perspective of set theory, both objects are sets of elements from the first level and therefore, morphisms are elements of the second level. This can be considered as a highly efficient way to deal with complexity by lifting the level of abstraction. Such an argument would have been convincing, if we had only two or, at most, a few levels of the hierarchy. However, the hierarchy is infinite, so lifting or lowering by one level is irrelevant. This does not preclude the fact that the category theory and its easy diagrammatic representation may have multiple good contributions to mathematics and its applications.

Funding: This research received no external funding.

Acknowledgments: The author would like to express his gratitude for the encouragement, criticism, and helpful comments from the reviewers and the editor.

Conflicts of Interest: The author declares no conflict of interest.

References

1. Dodig-Crnkovic, G.; Schroeder, M.J. (Eds.) *Contemporary Natural Philosophy and Philosophies, Part 1*; MDPI: Basel, Switzerland, 2019; Available online: https://www.mdpi.com/books/pdfview/book/1331 (accessed on 30 June 2020).
2. Dodig-Crnkovic, G.; Schroeder, M.J. Contemporary natural philosophy and philosophies. *Philosophies* **2018**, *3*, 42. [CrossRef]
3. Fallacies. *Internet Encyclopedia of Philosophy*; American Library Association: Washington, DC, USA, 1995; Available online: https://iep.utm.edu/fallacy/) (accessed on 16 August 2020).
4. Snow, C.P. *The Two Cultures*; Cambridge University Press: London, UK, 1959.
5. Conant, J.B. *General Education in a Free Society: Report of the Harvard Committee*; Harvard University Press: Cambridge, MA, USA, 1946.
6. Wilson, E.O. *Consilience: The Unity of Knowledge*; Vintage Books: New York, NY, USA, 1999.
7. Wilson, E.O. *The Meaning of Human Existence*; Liveright Publishing: New York, NY, USA, 2014.
8. Bateson, G. *Mind and Nature: A Necessary Unity*; E.P. Dutton: New York, NY, USA, 1979.

9. Butler, D. Theses spark twin dilemma for physicists. *Nature* **2002**, *420*, 5. [CrossRef] [PubMed]
10. Wilczek, F. Physics in 100 Years. Available online: http://frankwilczek.com/2015/physicsOneHundredYears03.pdf (accessed on 17 August 2020).
11. Simeonov, P.L.; Smith, L.S.; Ehresmann, A.C. Stepping Beyond the Newtonian Paradigm in Biology: Towards an Integrable Model of Life–accelerating discovery in the biological foundations of science, INBIOSA White Paper. In *Integral Biomathics: Tracing the Road to Reality*; Simeonov, P.L., Smith, L.S., Ehresmann, A.C., Eds.; Springer: Berlin/Heidelberg, Germany, 2012; pp. 319–418.
12. Capra, F. *The Tao of Physics*; Shambhala Publications: Boulder, CO, USA, 1975.
13. Laplane, L.; Mantovani, P.; Adolphs, R.; Chang, H.; Mantovani, A.; McFall-Ngai, M.; Rovelli, C.; Sober, E.; Pradeu, T. Why science needs philosophy. *Proc. Natl. Acad. Sci. USA* **2019**, *116*, 3948–3952. [CrossRef]
14. Von Foerster, H. (Ed.) *Cybernetics of cybernetics: Or, the Control of Control and the Communication of Communication*, 2nd ed.; Future Systems: Minneapolis, MN, USA, 1995.
15. Umpleby, S.A. Second-order science: Logic, strategies, methods. *Construct. Found.* **2014**, *10*, 16–23.
16. Müller, K.H. *Second-order Science: The Revolution of Scientific Structures*; Echoraum: Wien, Austria, 2016.
17. Lissack, M. Second order science: Examining hidden presuppositions in the practice of science. *Found. Sci.* **2017**, *22*, 557–573. [CrossRef]
18. Quine, W.V.O. Mr. Strawson on logical theory. *Mind* **1953**, *62*, 433–451. [CrossRef]
19. Laudan, L. *Science and Relativism: Some Key Controversies in the Philosophy of Science*; The University of Chicago Press: Chicago, IL, USA, 1990.
20. Niiniluoto, I. *Critical Scientific Realism*; Clarendon Press: Oxford, UK, 1999.
21. Dummett, M. Realism. *Synthese* **1982**, *52*, 55–112. [CrossRef]
22. De Caro, M. Naturalism and Realism. In *A Companion to Naturalism*; do Carmo, J.S., Ed.; Dissertatio's Series of Philosophy; University of Pelotas: Pelotas, Brasil, 2016; pp. 182–195.
23. Schroeder, M.J. Invariance as a tool for ontology of information. *Information* **2016**, *7*, 11. Available online: https://www.mdpi.com/2078-2489/7/1/11 (accessed on 15 June 2020). [CrossRef]
24. Lyotard, J.-F. Theory and history of literature. In *The Postmodern Condition: A Report on Knowledge*; University of Minnesota Press: Minneapolis, MN, USA, 1984; Volume 10.
25. Anderson, P. *The Origins of Postmodernity*; Verso: London, UK, 1998; pp. 24–27.
26. Bacon, F. *Novum Organum or True Suggestions for the Interpretation of Nature*; Devey, J., Ed.; P.F. Colier & Son: New York, NY, USA, 1902; Available online: http://www.gutenberg.org/files/45988/45988-h/45988-h.htm (accessed on 7 June 2020).
27. Bacon, R. *Opus Majus of Roger Bacon: Part I*; Kessinger Publishing: Whitefish, MT, USA, 2002.
28. Whorf, B.L. *Language, Thought, and Reality*; Caroll, J.B., Ed.; MIT Press: Cambridge, MA, USA, 1956.
29. Newton, I. *Philosophiae Naturalis Principia Mathematica, General Scholium*, 3rd ed.; University of California Press: Berkeley, CA, USA, 1726/1999; ISBN 0-520-08817-4.
30. Losee, J. *A Historical Introduction to the Philosophy of Science*; Oxford Univ. Press: London, UK, 1972.
31. Adams, D. *The Hitchhiker's Guide to the Galaxy*; Pan Macmillan Books: London, UK, 1979.
32. Tegmark, M. Shut up and calculate. *arXiv* **2007**, arXiv:physics/0709.4024v1.
33. Schroeder, M.J. Crisis in science: In search for new theoretical foundations. *Progress Biophys. Mol. Biol.* **2013**, *113*, 25–32. [CrossRef] [PubMed]
34. Schroeder, M.J. Structures and their cryptomorphic manifestations: Searching for inquiry tools. In *Algebraic Systems, Logic, Language and Related Areas in Computer Science, RIMS Kokyuroku*; Adachi, T., Ed.; Kyoto Research Institute for Mathematical Sciences, Kyoto University: Kyoto, Japan, 2019; p. 10.
35. Schroeder, M.J. Equivalence, (crypto) morphism and other theoretical tools for the study of information. *Proceedings* **2020**, *47*, 12. [CrossRef]
36. Schroeder, M.J. Intelligent computing: Oxymoron? *Proceedings* **2020**, *47*, 31. [CrossRef]
37. Schroeder, M.J. Logico-algebraic structures for information integration in the brain. In *Algebras, Languages, Computation and Their Applications, RIMS Kokyuroku*; Shoji, K., Ed.; Research Institute for Mathematical Sciences, Kyoto University, Kyoto: Kyoto, Japan, 2007; Volume 1562, pp. 61–72. Available online: http://www.kurims.kyoto-u.ac.jp/~{}kyodo/kokyuroku/contents/1562.html (accessed on 20 June 2020).
38. Wigner, E. The unreasonable effectiveness of mathematics in the natural sciences. In *Communications in Pure and Applied Mathematics*; New York University: New York, NY, USA, 1960; Volume 13.
39. Weyl, H. *Symmetry*; Princeton University Press: Princeton, NJ, USA, 1952.

40. Anderson, P.W. More is different. *Science* **1972**, *177*, 393–396. [CrossRef] [PubMed]
41. Schroeder, M.J. Concept of information as a bridge between mind and brain. *Information* **2011**, *2*, 478–509. Available online: http://www.mdpi.com/2078-2489/2/3/478/ (accessed on 1 June 2020). [CrossRef]
42. Schroeder, M.J. Towards cyber-phenomenology: Aesthetics and natural computing in multi-level information systems. In *Recent Advances in Natural Computing*; Ser. Mathematics for Industry 9; Suzuki, Y., Hagiya, M., Eds.; Springer: Tokyo, Japan, 2015; pp. 69–86.
43. Tversky, A.; Kahneman, D. Extensional versus intuitive reasoning: The conjunction fallacy in probability judgment. *Psychol. Rev.* **1983**, *90*, 293–315. [CrossRef]
44. Suppes, P. *Introduction to Logic*; Van Nostrand: Princeton, NJ, USA, 1957.
45. Simon, H.A. The axiomatization of physical theories. *Philos. Sci.* **1970**, *37*, 16–26. [CrossRef]
46. Tarski, A. Some methodological investigations on the definability of concepts. In *Logic Semantics, and Metamathematics: Papers from 1923 to 1938*; The Clarendon Press: Oxford, UK, 1956.
47. Carroll, L. *Alice's Adventures in Wonderland*; Pan Macmillan Books: London, UK, 1865.
48. Kroeber, A.L. Culture: A critical review of concepts and definitions. In *Papers of the Peabody Museum of American Archaeology and Ethnology*; Harvard University, Peabody Museum of American Archaeology: Andover, MA, USA, 1952; Volume 42.
49. Lovejoy, A.O. Nature as aesthetic norm. *Mod. Lang. Notes* **1927**, *7*, 444–450. [CrossRef]
50. Lovejoy, A.O. Some meanings of 'nature'. In *A Documentary History of Primitivism and Related Ideas*; Lovejoy, A.O., Boas, G., Eds.; Johns Hopkins Press: Baltimore, MD, USA, 1935; pp. 447–456.
51. Williams, R. Ideas of nature. In *Problems in Materialism and Culture: Selected Essays*; Williams, R., Ed.; Verso: London, UK, 1980; pp. 67–85.
52. Williams, R. *Keywords: A Vocabulary of Culture and Society*, rev. ed.; Bennet, T., Grossberg, L., Morris, M., Eds.; Blackwell: Malden, MA, USA, 2005; pp. 235–238.
53. Ogden, C.K.; Richards, I.A. *The Meaning of Meaning: A Study of the Influence of Language Upon Thought and of the Science of Symbolism*; Harcourt Brace Jovanovich: San Diego, CA, USA, 1989.
54. Shannon, E.C.; Weaver, W. *The Mathematical Theory of Communication*; University of Illinois Press: Urbana, IL, USA, 1949.
55. Kosso, P. Science and Objectivity. *J. Philos.* **1989**, *86*, 245–257. [CrossRef]
56. Schroeder, M.J. Hierarchic information systems in a search for methods to transcend limitations of complexity. *Philosophies* **2016**, *1*, 1–14. Available online: http://www.mdpi.com/2409-9287/1/1/1 (accessed on 15 June 2020). [CrossRef]
57. Schroeder, M.J. Exploring meta-symmetry for configurations in closure spaces. In *Developments of Language, Logic, Algebraic System and Computer Science, RIMS Kokyuroku*; Horiuchi, K., Ed.; Kyoto Research Institute for Mathematical Sciences, Kyoto University: Kyoto, Japan, 2017; pp. 35–42. Available online: http://www.kurims.kyoto-u.ac.jp/~{}kyodo/kokyuroku/contents/pdf/2051-07.pdf (accessed on 1 June 2020).
58. Worrall, J. Structural Realism: The best of both worlds? *Dialectica* **1989**, *43*, 99–124. [CrossRef]

MDPI
St. Alban-Anlage 66
4052 Basel
Switzerland
Tel. +41 61 683 77 34
Fax +41 61 302 89 18
www.mdpi.com

Philosophies Editorial Office
E-mail: philosophies@mdpi.com
www.mdpi.com/journal/philosophies